Shaping the Future: How Superlattices Revolutionize Material Design and Devices

Khanam

Table of Contents

Chapter 1. Superlattices and band structure engineering

1.1 Superlattices and photonic crystals

The concept of "superlattice" was first developed by L.Esaki and R.Tsu in 1970 by

considering a one-dimensional periodic potential within a 3D semiconductor system [1].

Specifically, they considered a superlattice formed by a periodic variation of alloy

composition or of impurity density introduced during epitaxial growth of semiconductors like

Ge, Si, Ge-Si alloys and GaAs. When the period of the superlattice is on the order of ~100 Å

(which is about 20 times the lattice constant of the host crystal) and is shorter than the

electron mean free path, the band structure of the electron is divided into a series of allowed

and forbidden minibands, with negative differential conductivity possible near the top edges

of the allowed bands. Shortly later, Esaki and Chang demonstrated transport signatures from

quantum states in an 85Å period GaAs-AlAs superlattice device. [2] Since then the concept

of superlattice has been extended from 1D superlattice (1DSL) to 2D[3, 4], 3D[5], or even

non-periodic superlattices[6], and the host crystals range from semiconductors[1, 2],

2DEGs[4, 7], and later 2D materials[8, 9, 10, 11, 12], resulting in exotic engineered band

structure features including satellite Dirac cones[10, 11, 12] and flat bands related to

superconductivity and Mott insulators[13, 14], Chapters 1~4 of this dissertation discuss

further efforts to engineer transport anisotropy and Dirac cone flattening/unflattening cycles

in a graphene 1D superlattice system.

A closely related concept is photonic crystals, which are periodic nanostructures engineered

for the purpose of controlling the motion of photons. Although photonic crystals occur

naturally in opals[15] and certain butterfly wings [16]; have long been studied and used in the

form of diffraction gratings (an example of a 1D photonic crystal), the term "photonic

crystal" was first coined by Eli Yablonovitch [17]and Sajeev John[18] in 1987 as they

proposed a 3D photonic crystal that exhibits a full photonic band gap. The concept of photonic crystals has recently been extended to surface plasmon polaritons (SPPs)[19], which are hybrid excitations of infrared photons and Dirac electrons with wavelengths much shorter than that of the incident photons. Just like electronic superlattices that may engineer novel band structure features not present in the host crystal, photonic crystals can tune the plasmonic band structure of SPPs hosted in graphene, leading the way to programmable SPP switches that will be further explored in Chapter 5.

1.2 Superlattices based on 2D materials

1.2.1 2D Materials and their heterostructures

In Richard Feynman's talk "There's Plenty of Room at the Bottom" in 1959, he envisioned a "great future" where "we can arrange all the atoms the way we want" and where there are "layered structures with just the right layers". [20]The latter has been at least partially realized with the advent of 2D materials and their heterostructures.

Graphene is the first 2D material to be unambiguously produced and identified[21]. It is a two-dimensional sheet of carbon atoms arranged in a honeycomb pattern. Tight-binding calculations show that graphene hosts Dirac points in its band structure where the energy dispersion is linear, indicating a carrier effective mass of zero. This is a crucial difference between graphene and semiconductor based two-dimensional electron gas (2DEGs), where the carriers have a parabolic band structure and therefore nonzero effective mass. As a result, graphene-based 1DSL systems feature flattening/unflattening cycles of their Dirac cones as superlattice modulation increases, which are not seen in 2DEG-based 1DSLs. (Chapters 2 and 4)

Numerous other classes of 2D materials, with one or few layers of atoms, have been discovered since the advent of graphene in 2004: hexagonal boron nitride (hBN), transition metal dichalcogenides (TMDCs), 2D magnets (CrI_3,$Cr_2Ge_2Te_6$) [22], and phosphorene [23].

They, along with graphene, can be stacked on each other thanks to interlayer van der Waals interactions, forming 2D heterostructures, or Feynman's "layered structures with just the right layers". Some of these heterostructures, such as magic angle twisted bilayer graphene and graphene-hBN moiré structures, are themselves interesting superlattice systems without the need of imposing further superlattice potentials. In my studies on the transport and optical properties of graphene subject to patterned superlattice potential detailed in this dissertation, hBN-graphene-hBN heterostructures are used to enhance carrier mobilities of graphene[24] and facilitate carrier density tuning by field effect, as hBN functions as a dielectric between graphene and device gates.

1.2.2 Moiré superlattices

A moiré pattern can be formed by a nonzero twisted angle or the mismatch in lattice constants between two adjacent layers of 2D materials.

The first moiré superlattice device based on 2D materials was made by Li et al[8] where graphene is placed on monolayer graphite at a nonzero twist angle, resulting in extra Dirac points that produce dI/dV peaks.

The twisted bilayer graphene (tBLG) band structure changes dramatically as a function of twist angle, and at the so-called "magic angle" of $1.1°$ a pair of isolated flat bands appear near zero Fermi energy, leading the way to Mott-like insulating behavior and superconductivity near half-filling of the flat band. [13]A similar effect was later discovered in twisted bilayer WSe_2 with transport signatures of correlated states seen at flat band half filling over a relatively wider range of twist angles from $4°$ to $5.1°$[14]..

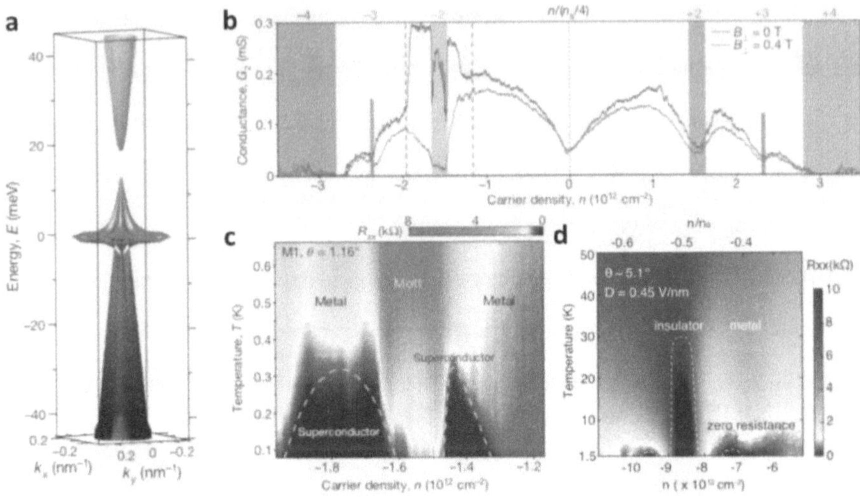

Figure 1.1 Twisted bilayer graphene and twisted WSe₂ moiré superlattices (a) Reprinted from [13] showing the band structure of tBLG at twist angle $\theta = 1.1°$, the "magic angle" where flat bands forms at zero Fermi energy (b) Reprinted from [13] showing the conductance of a $\theta = 1.16°$ device measured at $T = 70mK$. (c) Reprinted from [13] showing the longitudinal resistance R_{xx} as a function of temperature and carrier density, for the same $\theta = 1.16°$ device as in (b). At the half-filling of the valence flat band ($\frac{n}{\frac{n_s}{4}} = -2$) Mott-like insulating behavior and superconductivity are seen. (d) Reprinted from [14] showing the longitudinal resistance R_{xx} as a function of temperature and carrier density, for a $\theta = 5.1°$ twisted WSe₂ device at displacement field $D = 0.45V/nm$. It shows Mott-like insulating behavior and superconductivity similar to those in tBLG.

The first graphene/hexagonal boron nitride (hBN) moiré superlattice device was made by Xue et al in a graphene-hBN system in 2011[9]. By superposing graphene and hBN lattices on each other at a twist angle of near zero, a moiré pattern is formed with wavelength on the order of 10nm, about 40 times those of the BLG or hBN crystals. The existence of the moiré pattern can be confirmed by scanning tunneling microscopy (STM). One year later,

Yankowitz et al confirmed that the moiré pattern acts as a superlattice modulation potential,

as measurement of dI/dV shows satellite peaks at nonzero gate voltages that correspond to

nonzero carrier densities, indicating local density of states (LDOS) minima and the existence

of satellite Dirac cones.[10] Later, Dean et al discovered similar effects in a bilayer graphene

(BLG)-hBN system by demonstrating the existence of satellite longitudinal resistance

peaks.[11]

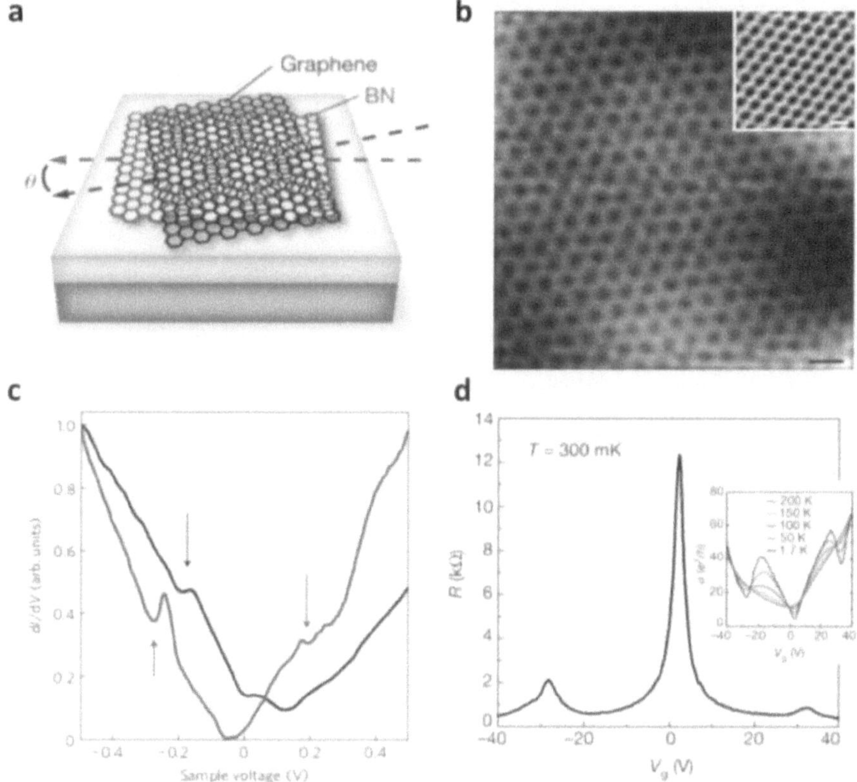

Figure 1.2 hBN/Graphene moiré superlattices (a) Reprinted from [11]. A sketch showing

the emergence of a hBN/Graphene moiré superlattice. The moiré wavelength depends on the

twist angle θ. (b) Reprinted from [9]. An STM topographic image of a hBN/Graphene moiré superlattice. Scale bar: 2μm. Inset is a zoomed in view with a scale bar of 0.3μm. (c) Reprinted from [10]. dI/dV of two hBN/graphene moiré samples, shown in red (13.4nm wavelength) and **black** (9.0nm wavelength) respectively, as a function of sample voltage, which determines the carrier density in graphene. Arrows indicate dips in dI/dV curves that correspond to the emergence of satellite Dirac cones. (d) Reprinted from [11]. Longitudinal resistance in a hBN/bilayer graphene device at $T = 0.3K$ as a function of gate bias that determines the carrier density in graphene. Inset shows the change of longitudinal conductivity versus temperature, indicating that the satellite features disappear at $T > 100K$.

Although moiré superlattices have enabled these remarkable feats of band structure engineering, the geometries of moiré patterns, and thus of the superlattice modulation potentials, are limited to triangular and certain quasicrystal patterns with wavelengths ~10nm. In addition, the physical nature of moiré superlattice modulation is a complicated mixture of electrostatic and strain modulations [25], with no easy way to separate contributions of these two effects. Furthermore, the strength of superlattice modulation cannot be independently tuned. These restrictions severely limit the application of moiré superlattices in engineering electronic structures.

1.2.3 Patterned superlattices

The idea of patterned superlattices is to directly pattern the desired geometry of the superlattice modulation onto the device using nanolithography techniques. Patterned superlattices is an established technique in the field of 2DEG superlattice research but remains a burgeoning subfield in 2D material superlattices. Due to the relative ease of making electrical contacts to graphene than TMDCs, all the experimental work on patterned superlattices in 2D materials so far are made on graphene.

There are three main routes to patterned superlattices: Patterning the 2D material itself, gate patterning and dielectric patterning. In 2019, Jessen et al directly dug out a 35nm-pitch triangular hole pattern out of a graphene sheet encapsulated between two hBN layers.[26] Despite the direct patterning on graphene, their device still has a carrier mobility around 800 cm^{-2} V^{-1} s^{-1}, ensuring ballistic transport on the scale of the minimum feature size of ~10nm. Multiple resistance peaks unseen in pristine graphene are seen in this nanopatterned graphene device, and they diminish as temperature increases, suggesting modifications to the graphene band structure. Furthermore, signatures of semiclassical "skipping orbit states" around the etched hole sites are extracted from magnetotransport data.

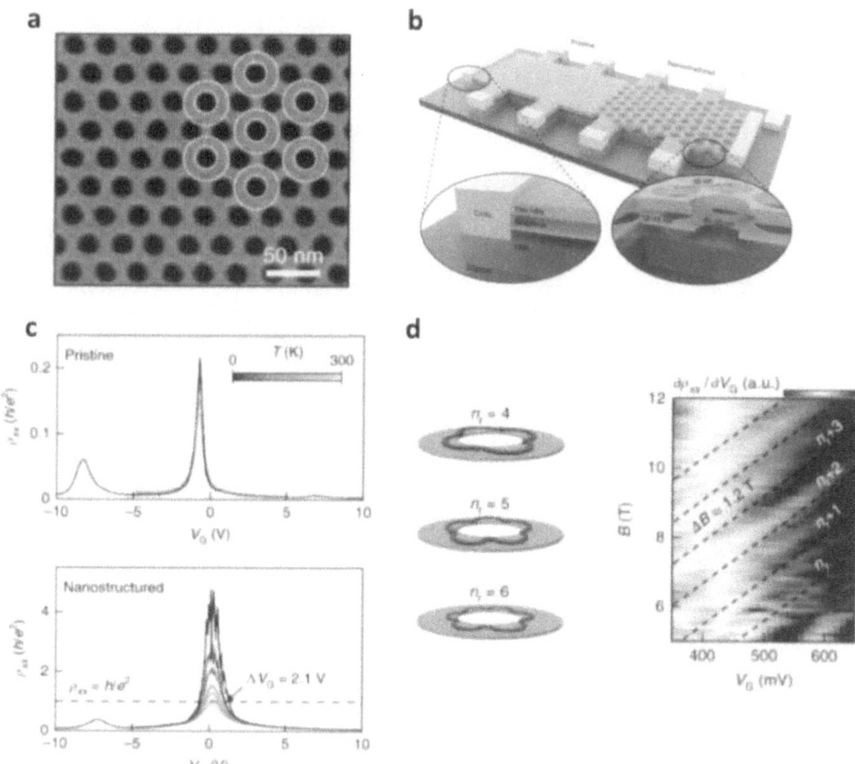

Figure 1.3 Lithographically patterned graphene Reprinted from [26]. (a) SEM image of a hBN-graphene-hBN heterostructure patterned by etching away the graphene (and the hBN on top of it) in a triangular array of holes with period 35nm. Minimum feature size = 12~15nm. (b) Schematics of a Hall bar device used for transport measurement. Data from pristine graphene and patterned graphene can be measured simultaneously. (c) Longitudinal resistivity versus gate voltage, thus the carrier density, at various temperatures. At low temperature nanostructured graphene exhibits resistivity peaks that are smoothed out at high temperature. The side peaks at $V_G = \pm 7.6V$ are attributed to moiré superlattice. (d) In a semiclassical picture, under finite magnetic field the charge carriers form "skipping orbits" around the perimeters of the etched holes. Magnetoresistance therefore is expected to show oscillation with a period $\Delta B = 8\hbar/e(2\pi R_1)^2$ where R_1 is the hole radius. This oscillation is confirmed with $\partial\rho_{xx}/\partial V_G$ data.

The direct patterning on graphene, while allowing for arbitrary geometries of superlattice modulation, is unable to provide a knob to tuning the strength of modulation once the device is fabricated. Since parts of the graphene itself is etched, the Hamiltonian of the patterned graphene system is very different from that of graphene (as can be corroborated by the presence of "skipping orbit states"), and cannot be understood as simply adding a position dependent superlattice modulation potential term to the pristine graphene Hamiltonian. Both gate patterning and dielectric patterning preserve the graphene itself, while electrostatically inducing a carrier density distribution pattern $n(\vec{r})$ on graphene surface with a pitch of ~40nm. The induced carrier density can be broken into two parts, $n(\vec{r}) = \bar{n} + \Delta n(\vec{r})$, where $\bar{n}$ is the average carrier density in graphene and $\Delta n(\vec{r})$ is the periodic modulation. $\Delta n(\vec{r})$ results in a local variation in Dirac point energies $E_D\big(n(\vec{r})\big) = -sgn\big(n(\vec{r})\big)\hbar v_F\sqrt{\pi|n(\vec{r})|}$ where v_F is the Fermi velocity of pristine graphene. The

Hamiltonian of the modulated graphene system can thus be expressed as the sum of a pristine graphene Hamiltonian term $H_{graphene} = \pm \hbar v_F \boldsymbol{p} \cdot \boldsymbol{\sigma}$ and the effective superlattice potential term $U_{SL}(\vec{r}) = E_D\big(n(\vec{r})\big) - E_D(\bar{n})$.

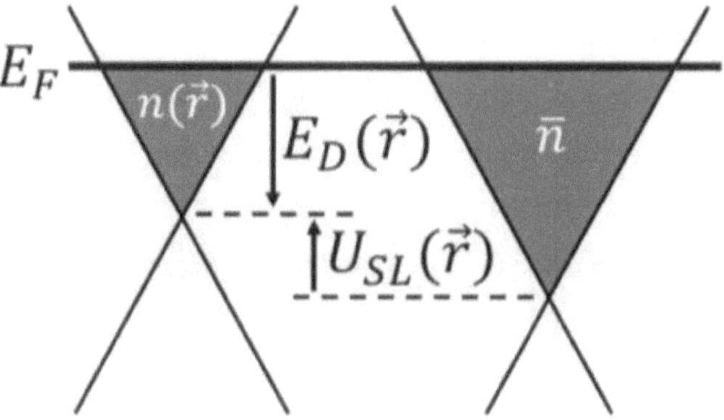

Figure 1.4 Origin of the effective superlattice potential $\boldsymbol{U_{SL}}$ Reprinted from Dr. Carlos Forsythe's dissertation *Fractal Hofstadter Band Structure in Patterned Dielectric Superlattice Graphene Systems*, Fig. 3.11.

For both gate patterning and dielectric patterning schemes, two electrostatics gates are necessary to tune two parameters, namely the average carrier density in graphene $\bar{n}$ and superlattice potential U_{SL}. In the case of gate patterning, one of the gates (called the SL gate) is nanopatterned, and the application of gate voltages on this gate induces $\Delta n(\vec{r})$. In the case of dielectric patterning, a nanopattern is etched onto the dielectric materials (SiO_2/hBN/hydrogen silsesquioxane, etc..) that separate graphene and one of the two gates (also called the SL gate). Air takes over the etched sites, resulting in a local dielectric constant inhomogeneity that translates into $\Delta n(\vec{r})$ at nonzero SL gate voltages. Notice that in both schemes, although the SL gate itself mostly dictates the strength of superlattice modulation U_{SL}, by the parallel capacitor model both the SL gate and the non-SL gate contribute to the average carrier density $\bar{n}$ of the grounded graphene sheet. $\bar{n}$ is therefore a linear combination of the voltages of both gates.

Gate patterning and dielectric patterning are able to achieve superlattice modulation on the order of tens of meV, for a 40nm-pitch 2D superlattice on graphene, while maintaining relatively high carrier mobilities compared to directly patterning graphene. The relative strengths of these two schemes are still up to debate. More in-depth discussions on electrostatically engineered graphene 2D superlattices will be continued in Section 1.3.

1.3 Graphene under electrostatically engineered 2D superlattice potentials

In a 2018 paper Forsythe et al investigated a graphene device subject to a 40nm-pitch triangular superlattice potential made by dielectric patterning. [12] Transport data was taken at temperature T=220~250 mK. Besides a main longitudinal resistance peak at the charge neutrality point (CNP), when the superlattice potential is turned on by setting $V_{SL} = \pm 50V$ three pairs of side resistance peaks appear at carrier densities $n/n_0 = \pm 4, \pm 7, \pm 12$ respectively. (n_0 is the carrier density at which there is one electron per superlattice unit cell) These side resistance peaks disappear when the superlattice modulation is turned off with $V_{SL} = 0V$. Plotting resistance as a function of V_{SL} and carrier density n also show blobs of resistance maxima in addition to the $n = 0$ blob that already exists in pristine graphene. These blobs happen at fixed carrier densities $n/n_0 = \pm 4$, showing that V_{SL}, which determines the strength of superlattice modulation, has no effect on the carrier densities at which the side resistance peaks occur.

The presence of these extra resistance maxima can be explained by the band structure of the graphene 2DSL system. The cone-shaped pristine graphene band structure is folded at the edges of the superlattice 1st Brillouin zone, forming superlattice minibands, and resulting in density of states (DOS) minima and thus resistance maxima. The 1st miniband, colored yellow in Fig. 1.5c, covers almost all the superlattice 1st Brillouin zone. Given the 4-fold degeneracy in graphene, two from spin and two from valley, the upper edge of the yellow band happens at $n/n_0 = 4$. The $n/n_0 = 12$ resistance maximum correspond to DOS minimum at the upper edge of the 3rd miniband and the lower edge of the 4th miniband. There is significant overlap between the 2nd and 3rd minibands, nevertheless DOS maxima exist at $n/n_0 = 6 \sim 7$, explaining the resistance maxima at $n/n_0 = \pm 7$.

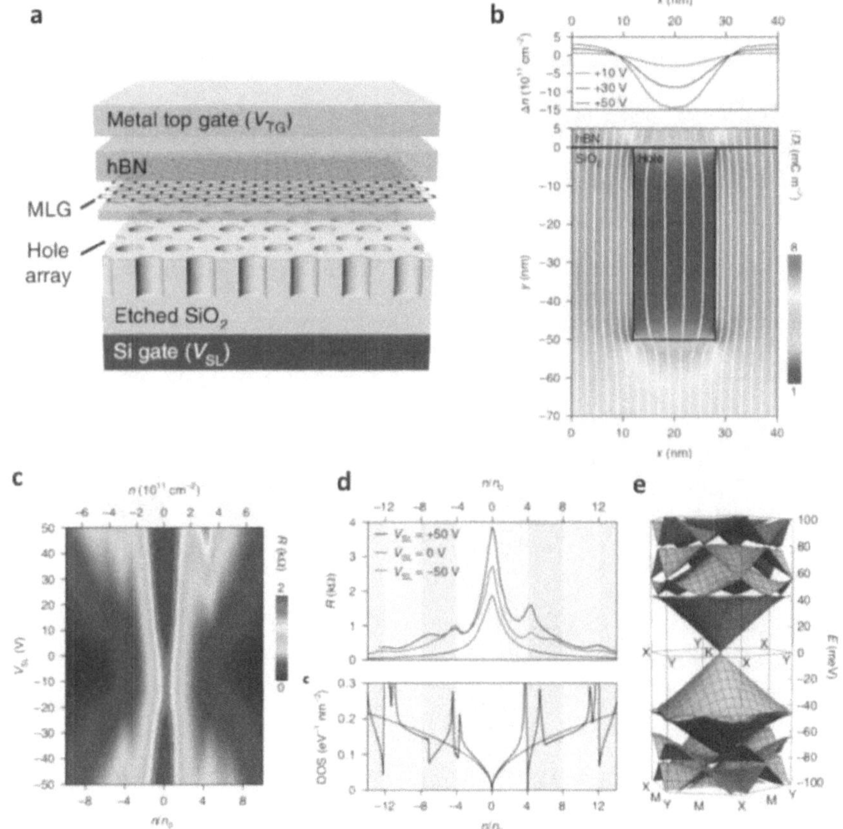

Figure 1.5 Graphene under patterned dielectric 2DSL Reprinted from [12]. (a) Schematics of a device. A triangular hole array with pitch 40nm is etched on SiO_2. The strength of the superlattice modulation is mostly controlled by the Si gate, while both the metal gate and Si gate control the carrier density. (b) Simulation of electric field and carrier

density modulation Δn at nonzero Si gate bias. (c) Longitudinal resistance as a function of Si gate bias and carrier density. (d) Measured longitudinal resistance at three different Si gate biases. The locations of the side resistance peaks match the locations of DOS minima. (e) Calculated band structure of the graphene 2DSL system with $V_{SL} = 50V$. The miniband colors in (e) match the colors in (d).

Under finite magnetic field perpendicular to the sample surface, graphene develops Landau levels that exhibit as Landau fans emanating from CNP in longitudinal magnetoresistance data versus magnetic field and carrier density. In addition to this main fan, two side fans emanate from $n/n_0 = \pm4$, as mini-Dirac cones formed by the 1st and 2nd minibands develop their own Landau levels. The interplay of magnetic field and a periodic structure provided by the superlattice give rise to fractal energy gaps called Hofstadter minigaps [27], and the existence of side Landau fans confirm the presence of such Hofstadter minigaps.

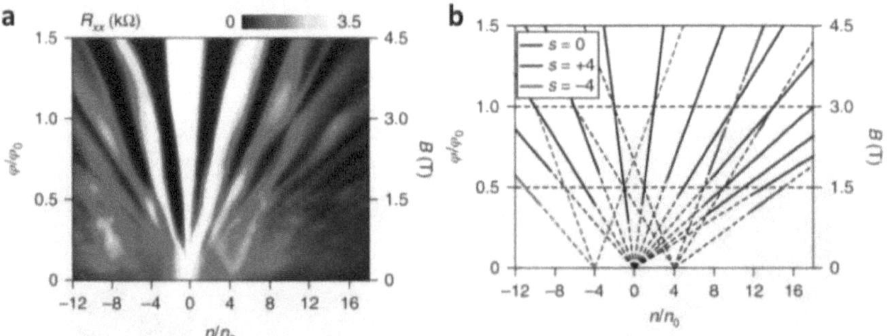

Figure 1.6 Magnetotransport of graphene under patterned dielectric 2DSL Reprinted from [12]. (a) Longitudinal resistance R_{xx} as a function of magnetic field, expressed in terms of both B and ϕ/ϕ_0, the number of flux quanta in a superlattice unit cell, and carrier density n/n_0, the number of electrons in a superlattice unit cell. $T \approx 250mK$. Hofstadter's fractal spectrum can be described by two topological integers (s,t) satisfying $n/n_0 = t(\phi/\phi_0) + s$. Besides the main Landau fan that already exists in pristine graphene ($s = 0$),

features related to side Landau fans ($s = \pm 4$) are also visible. (b) Tracing of magnetoresistance minima from plot (a). Hofstadter minigaps associated with satellite Dirac cones are colored in red ($s = -4$) and blue ($s = 4$) .

In 2020, Huber et al were able to replicate most of the results above by measuring a 40nm-pitch square superlattice graphene device made by gate patterning. [28] Their reported carrier mobility is 40,000 cm^{-2} V^{-1} s^{-1}, about one order of magnitude smaller than the 300,000 cm^{-2} V^{-1} s^{-1} result achieved in Forysthe's patterned dielectric device. In addition to side longitudinal resistance peaks at zero magnetic field and side Landau fans at finite magnetic fields already present in the data from the patterned dielectric device, it was shown experimentally that quantum Hall resistance can be a non-monotonic function of carrier density, unlike a monotonically decreasing sequence as expected in pristine graphene. The non-monotonicity of quantum Hall resistance can be explained by the competition between Hofstadter minigaps associated with different Landau fans, as a lower Landau level from an energetically higher Dirac cone ($s = +4,\ t = 6$ in Fig 1.7c) can have a higher energy than a higher Landau level from an energetically lower Dirac cone ($s = -4,\ t = 10$ in Fig 1.7c). This is a long-predicted property of Hofstadter's fractal energy spectra[29].

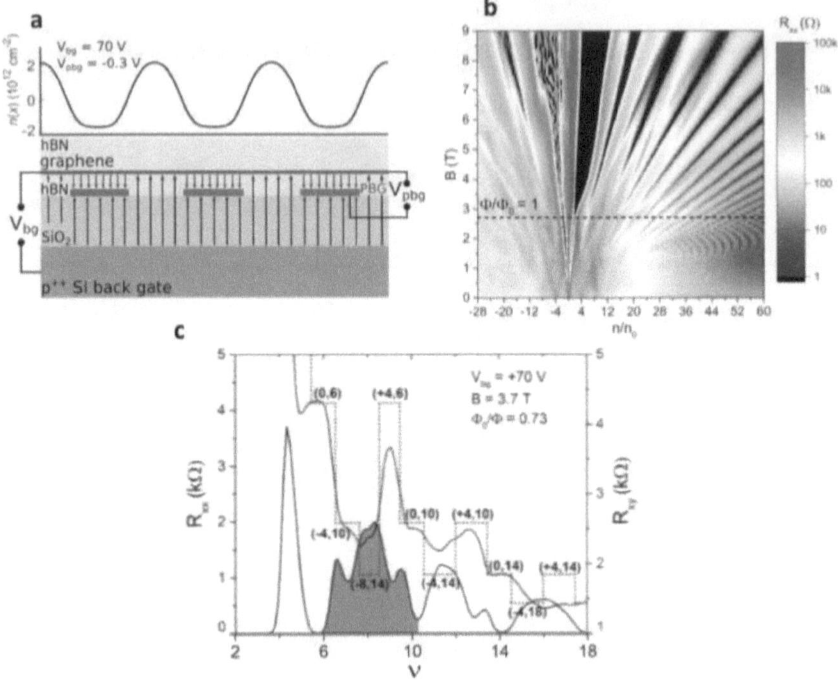

Figure 1.7 Magnetotransport of graphene under patterned gate 2DSL Reprinted from

[28]. (a) Device schematics. One of two back gates of this device is patterned with a

superlattice. Compare with the patterned dielectric scheme in Fig. 1.6(a). (b) Longitudinal

magnetoresistance as a function of magnetic field and carrier density. T=1.5K. Side fan

features related to $s = \pm 4$ and $s = -8$ can be resolved. (c) Hall resistance non-

monotonicity at $V_{bg} = 70V, B = 3.7T$. Blue line is the measured Hall resistance R_{xy} as a

function of electron filling factor $\nu = nh/eB$. Red dotted line shows the ideal nonmonotonic

quantum Hall sequence.

Graphene 2DSL systems are the first steps towards artificial band structure engineering in

graphene, but at zero magnetic field there are few significant novel features other than satellite Dirac cones. The strength of superlattice modulation also does not have a significant effect on the shape of the band structure. Owing to the dimensionality mismatch between a 1D superlattice and a 2D material in graphene, more exciting possibilities await in the quest for band structure engineering.

1.4 Summary

Superlattices can engineer electronic or optical/plasmonic band structure in 2D materials. In the latter case superlattices are often referred as photonic crystals. Both are explored in this dissertation.

Superlattice hosted on 2D materials can be formed by moiré patterns that naturally occur in neighboring 2D layers with different lattice constant or nonzero twist angle, or by artificial patterning. Artificial superlattice patterning is a more versatile approach of band structure engineering, with more control on geometry and modulation strength. Among the three artificial patterning methods, gate patterning and dielectric patterning does not invade the 2D material itself and provide higher carrier mobilities.

Graphene subject to patterned 2D superlattice potential develop additional Dirac cones at zero magnetic field and Hofstadter minigaps at finite magnetic field. But there are even more exotic opportunities of band structure engineering.

Chapter 2. Band structure of a graphene 1D superlattice system

2.1 Dimensional mismatch

Graphene 1D superlattices provide more opportunities of band structure engineering than graphene 2D superlattices, owing to the fact that graphene is a 2D material. The difference between 1D and 2D gives rise to two unique aspects of graphene 1DSL not found in graphene 2DSL: transport anisotropy and heavy dependence of band structure shape (and thus carrier densities of resistance maxima) on the strength of superlattice modulation.

1DSL can be described as an effective superlattice potential $U_{SL}(\vec{r})$ that depend on only one of the two spatial variables allowed in the graphene plane, which we call x by convention. Experimentally 1DSL can be produced by etching a series of equally spaced parallel lines on a gate or on the dielectric. It is easy to imagine that the band structure of graphene 1DSL would be highly anisotropic and may lead to difference in transport behaviors along x and y directions. As it will turn out, even the main Dirac point at CNP would be highly anisotropic under certain strengths of SL modulation, creating resistance maxima in the x direction while minima in the y direction.

In 2DSL the strength of SL modulation has limited effects on the shape of the band structure. For 2nd miniband and higher, stronger SL modulation tends to push neighboring minibands apart, reducing possible overlapping. (The overlapping between 2nd and 3rd miniband is responsible for the resistance peak at $n/n_0 = \pm 7$ in the graphene 2DSL by Forsythe et al, without which the peak should occur at $n/n_0 = \pm 8$). But the 1st miniband always start from the K point in the k-space at CNP as a cone and touches all edges of the SL 1st Brillouin zone, regardless of SL modulation strength. Therefore the 1st miniband always covers most of the SL 1st Brillouin zone, guaranteeing $n/n_0 = \pm 4$ resistance maxima[1] (See Fig. 1.5c)

There is no such guarantee in graphene 1DSL, as the SL 1st Brillouin zone has dimensions $\frac{2\pi}{L} \times \frac{2\pi}{a_{graphene}}$, where L, the 1DSL lattice constant, is about 50nm and the lattice constant of graphene is only 0.246nm! Therefore, effectively the SL 1st Brillouin zone has size $\frac{2\pi}{L} \times \infty$

and the 1st miniband, or any miniband, has no hope of covering it completely, and there is no

equivalent "$n/n_0 = \pm 4$ resistance maxima" argument in graphene 1DSL. In fact, the carrier

densities of all the resistance maxima would change as SL modulation changes.

So here is the bigger question: Does the dimensionally mismatch in 1DSL or 2DSL systems

hosted in 3D materials result in even more exciting band structure engineering possibilities?

2.2 The Hamiltonian

Park et al [2] first investigated the band structure of a graphene 1DSL system in 2008. More

detailed theoretical studies on this system were later done by Barbier et al[3]. Here we

describe the Hamiltonian of the system, using notations from [3], as $H = \pm \hbar v_F \boldsymbol{k} \cdot \boldsymbol{\sigma} + \mathbb{I}V$

where $\boldsymbol{\sigma}$ is the Pauli matrix vector, $\boldsymbol{k} = (k_x, k_y)$ is the momentum measured from K-point

of the graphene Brillouin zone, $\mathbb{I}$ is the 2x2 identity matrix. V(x) is a 1D square-wave

potential satisfying $V(x + L) = V(x)$. Each period consists of a "well" region and a

"barrier" region with width W_w and W_b respectively. The well region has V(x)=0 while the

barrier region has $V(x) = V_0 > 0$. The task is therefore solving the Schrödinger Equation

$$H\Psi = E\Psi \text{ with } H = \begin{pmatrix} V & -iv_F\hbar(\partial_x - i\partial_y) \\ -iv_F\hbar(\partial_x + i\partial_y) & V \end{pmatrix}$$

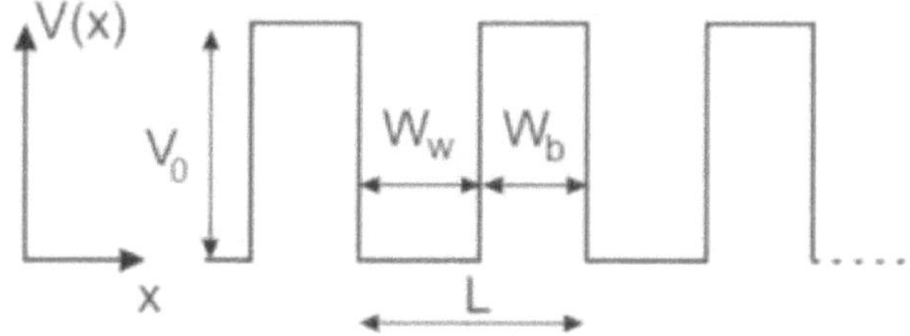

Figure 2.1 Schematics of the superlattice V(x) and the definitions of L, W_w, W_b, V_0

Reprinted from [3]

We hereby define the following dimensionless variables: $\epsilon = \frac{EL}{v_F \hbar}$, $u(x) = \frac{V(x)L}{v_F \hbar}$, $u = \frac{V_0 L}{v_F \hbar}$,

$s(x) = \text{sgn}[\epsilon - u(x)]$, $\lambda(x) = \left[(\epsilon - u(x))^2 - L^2 k_y^2 \right]^{\frac{1}{2}}$, $\tan \phi(x) = \frac{L k_y}{\lambda(x)}$, $\lambda =$

$\left[\left(\epsilon + \frac{u W_b}{L} \right)^2 - L^2 k_y^2 \right]^{1/2}$, $\Lambda = \left[\left(\epsilon - \frac{u W_w}{L} \right)^2 - L^2 k_y^2 \right]^{1/2}$, $G = \left[\left(\epsilon + \frac{u W_b}{L} \right) \left(\epsilon - \frac{u W_w}{L} \right) - \right.$

$\left. L^2 k_y^2 \right] / \lambda \Lambda$

Then eigenstates of the Hamiltonian H have the form $\psi(x) e^{ik_y y}$ with

$$\psi(x) = \begin{pmatrix} 1 \\ s(x) e^{i\phi(x)} \end{pmatrix} e^{i\lambda(x)x/L}, \psi(x) = \begin{pmatrix} 1 \\ -s(x) e^{-i\phi(x)} \end{pmatrix} e^{-i\lambda(x)x/L} \tag{1}$$

And the eigenenergies ϵ at a given (k_x, k_y) can be found by solving the following

equation: ([3], Eq. 7)

$$\cos k_x L = \cos \left(\frac{\lambda W_w}{L} \right) \cos \left(\frac{\Lambda W_b}{L} \right) - G \sin \left(\frac{\lambda W_w}{L} \right) \sin \left(\frac{\Lambda W_b}{L} \right) \tag{2}$$

Detailed discussions on the solution of (2) will be the focus of the remainder of this chapter.

Before doing this, we note that the dimensionless strength of SL modulation $u = \frac{V_0 L}{v_F \hbar}$ is

proportional to the SL pitch L. In a patterned dielectric superlattice device, the maximum

value of effective SL potential in the barrier regions, $V_{0,max}$, is mostly determined by the

breakdown voltage of the patterned dielectric, with very little dependence on L and the

depth of etched SL trenches once the depth reaches about 10nm. To be able to see more

features of graphene 1DSL in transport data, we need a higher u_{max} and therefore a higher

SL pitch L. This is very different from the case in graphene 2DSL where we strive for a pitch

as small as allowed by the e-beam lithography system in order to increase the carrier density

at which the first satellite Dirac cones appear at $n/n_0 = 4$, lest the side resistance peaks get

mired in the width of the main peak.

2.3 The case of W_w=0.5L

2.3.1 Evolution of band structure as u changes

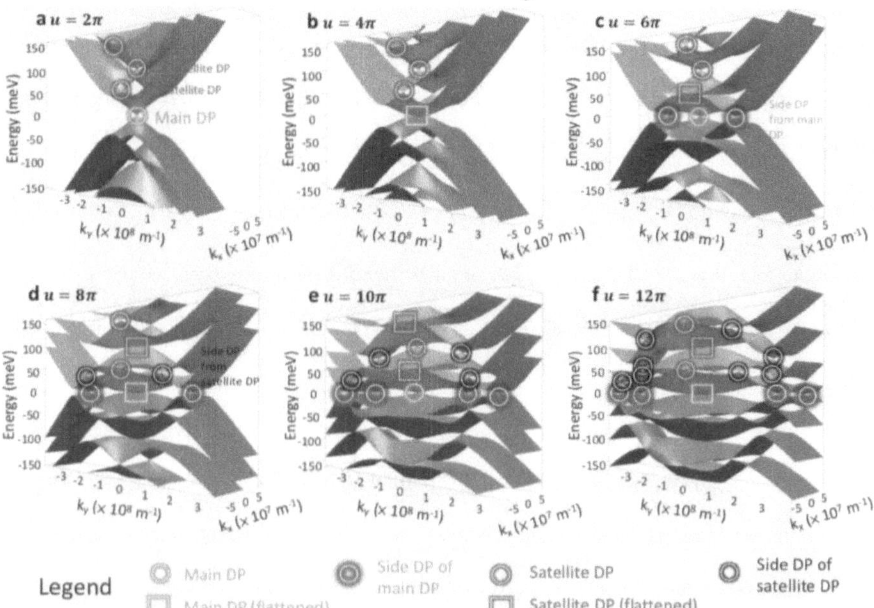

Figure 2.2 Calculated band structures for a $L = 55nm$, $W_w = 0.5L$ **graphene 1DSL system at six different strengths of SL modulation** u Dirac points that have linear dispersion relations in both k_x and k_y are called "unflattened" and labelled by circles. Other Dirac points have linear dispersions in k_x but higher order dispersion in k_y are called "partially flattened", or just "flattened" and labelled by rectangles. At small u certain higher order DPs are not very developed with nonlinear dispersion in k_y, but their mechanism and resulting transport signatures are different from those labelled by rectangles, so they are still labelled by circles.

We first consider the case of $W_w = 0.5L$, where the "well" and "barrier" regions have the same length. In this scenario the band structure is symmetric about E=0 so we need only consider the conduction bands.

Figure 2.2a shows graphene 1DSL band structure at $u = 2\pi$. Besides the main Dirac point labelled in green, a sequence of satellite DPs labelled in red appear at nonzero energies. The 1st, 3rd, 5th ... satellite DPs occur at $(k_x, k_y) = (0,0)$ while the main DP, as long as the 2nd, 4th, 6th... satellite DPs occur at $(k_x, k_y) = (\pm\frac{\pi}{L}, 0)$. This alternating of DP locations will turn out to hold true for all u. All the DPs, main or satellite, faithfully replicate the $E = \pm\hbar k v_F$ dispersion in the k_y direction but the Fermi velocity in the k_x direction is smaller than that of pristine graphene. (Fig. 2.3) For 2nd and higher satellite DPs, the Dirac point features are not fully developed, as there is very little energy difference between the upper and lower bands that constitute the DPs, to the point that these "Dirac points" are not very different from the simple band crossings between two Dirac cones centered at K and $K + nG$ ($n \geq 2$, G is the reciprocal lattice vector of the 1DSL), resulting in less conspicuous DOS minima than the main and 1st satellite DP.

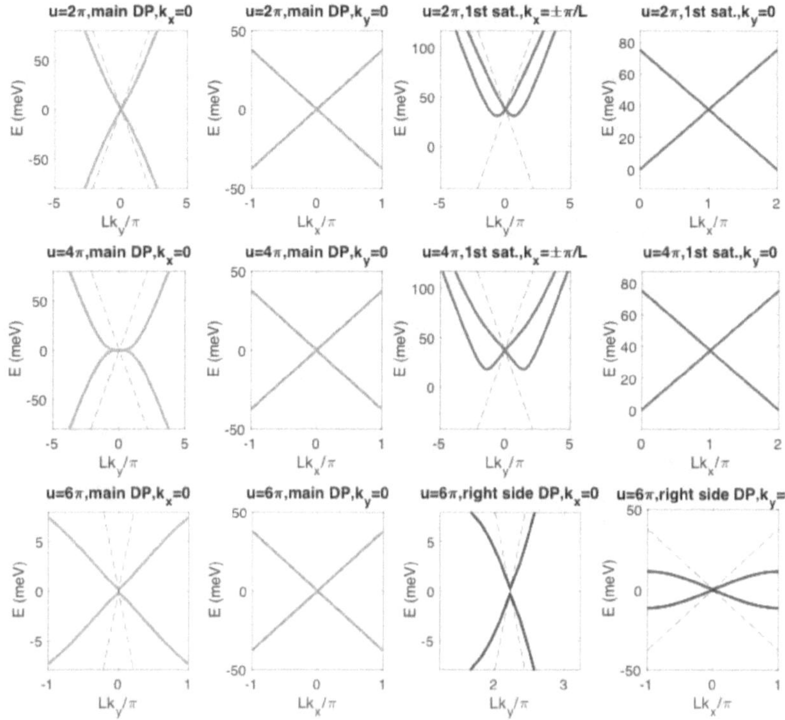

Figure 2.3 Band dispersion at certain Dirac points in graphene 1DSL Colors of the bands

correspond to the classification of DPs as explained in the legend in Figure 2.2. Dashed lines

indicate the dispersion relation of pristine graphene $E = \pm \hbar k v_F$.

Figure 2.2b shows graphene 1DSL band structure at $u = 4\pi$. The most striking feature is the

anisotropic flattening of the main Dirac point, labelled by rectangles in Figure 2.2 and 2.5.

The dispersion relations become $E \sim k_y^3$ at small k_y [3] while preserving the pristine

graphene $E = \pm \hbar k v_F$ relation in the k_x direction. All the satellite Dirac points do not

show anisotropic flattening. In addition, the 2nd satellite DP becomes fully developed.

Figure 2.2c shows graphene 1DSL band structure at $u = 6\pi$. Now the main DP is back to a

cone shape again, but it is accompanied by two side DPs at $E = 0$, $k_x = 0, k_y \neq 0$. Side

DPs associated with the main DP are labelled by blue. The band dispersions of the main DPs, side DPs and pristine graphene are different from one another. The main DP preserves the pristine graphene dispersion in k_y but has a much smaller Dirac velocity in k_x. For the side DPs, the Dirac velocity in k_x and k_y are both smaller than v_F. (Figure 2.3) The renormalization of Dirac velocity in k_y is a hallmark of side DPs .The 1st satellite DP becomes anisotropically flattened in k_y. But unlike the case in the flattening of the main DP at $u = 4\pi$ the band structure is not symmetric in energy about the DP, with the upper band being "flatter" than the lower band.

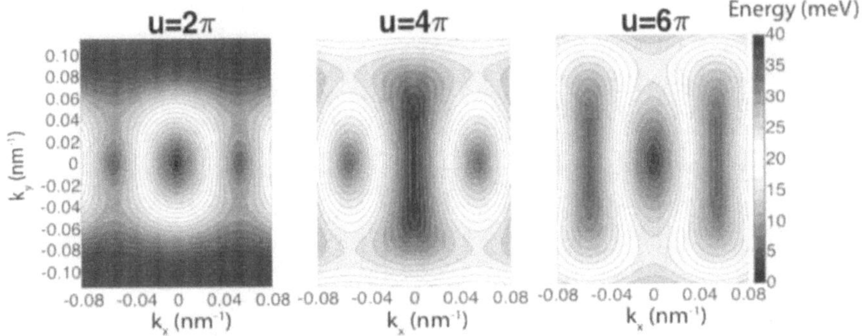

Figure 2.4 Band structure energy contours of graphene 1DSL at $u = 2\pi, 4\pi, 6\pi$ Notice that the k_x range shown here is wider than $2\pi/L$. The anisotropically flattened bands at $u = 4\pi, 6\pi$ can be easily identified. Also, it is worth noticing that the main and satellite DPs do not have isotropic dispersion seen in pristine graphene, as shown by the elliptical shape of the contour lines around the DPs. This plot also shows the existence of open orbits at high k_y along the k_x direction. The two side DPs at $u = 6\pi$ are not shown.

At $u = 8\pi$ the main DP is anisotropically flattened, but still accompanied by two side DPs like the case in $u = 6\pi$. The 1st satellite DP is unflattened and also accompanies by two side DPs labelled by **black** in Figure 2.2. Unlike the case of the "blue" side DPs accompanying the main DP, the "black" side DPs occur at a slightly lower energies than the satellite DP they accompany. The 2nd satellite DP is also being anisotropically flattened, with the upper band being flatter than the lower band, as we have seen in the 1st satellite DP at $u = 6\pi$. At $u = 10\pi$ we have 4 side DPs for the main DP which is unflattened, the 1st and 3rd satellite DPs are anisotropically flattened, the 1st and 2nd satellite DPs have 2 side DPs each. At $u = 12\pi$ we have 4 side DPs for the main DP. The main DP and $2^{nd}, 4^{th}$ satellite DPs are anisotropically flattened. The 1st satellite DP has 4 side DPs and the 2nd and 3rd satellite DPs have 2 side DPs each.

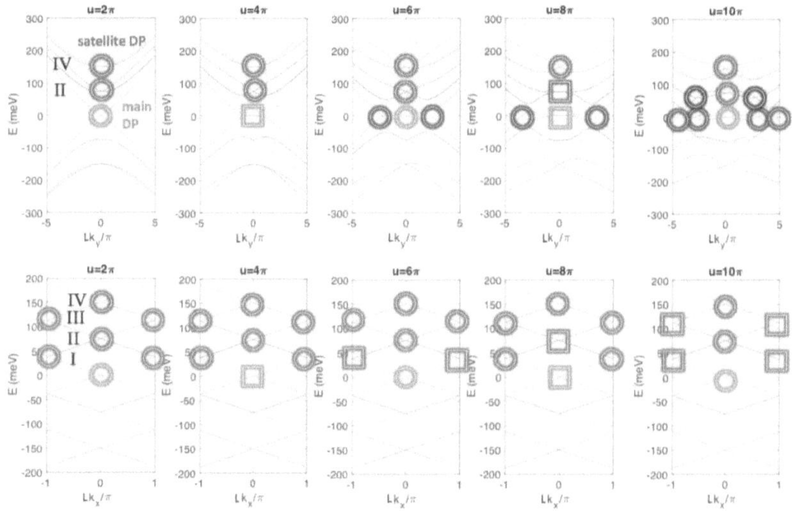

Figure 2.5 Slices of graphene 1DSL band structures at $u = 2\pi, 4\pi, 6\pi, 8\pi, 10\pi$ **along** $k_x = 0$ (top row) **and** $k_y = 0$ (bottom row) Dirac points are labelled by circles (unflattened) or rectangles (flattened in k_y). For the meaning of colors see legend in Figure 2.2. Roman numerals indicate the order of satellite DPs. Notice that not all DPs are shown, as the side DPs of odd-ordered satellite DPs have nonzero k_x and k_y. For example, the side DPs associated with the 1st order satellite DP at $u = 8\pi, 10\pi$, though shown in Figure 2.2, are not shown here

	Main DP	Satellite DP	Side DP of Main DP	**Side DP of Satellite DP**
k_x	0	$\pm\pi/L$ for odd l, 0 for even l	0	$\pm\pi/L$ for odd l, 0 for even l
k_y	0	0	$\neq 0$	$\neq 0$
Energy	0	$(l\pi)\dfrac{V_0 L}{\hbar v_F}$	0	$< (l\pi)\dfrac{V_0 L}{\hbar v_F}$
$\left\|\dfrac{\partial E}{\partial k_x}\right\|$	$= \hbar v_F$	$= \hbar v_F$	$< \hbar v_F$	$< \hbar v_F$
$\left\|\dfrac{\partial E}{\partial k_y}\right\|$	$< \hbar v_F$	$< \hbar v_F$	$< \hbar v_F$	$< \hbar v_F$
Anisotropic Flattening	May happen	May happen	No	No
Range of u that such DP exists	$[0, +\infty)$	$(0, +\infty)$	$(4\pi, +\infty)$	$(6\pi, +\infty)$

Table 2.1 Summary of the properties of the four types of DPs available in the graphene 1DSL system l refers to the order of the satellite DP and is assumed to be positive.

The anisotropic flattening of main and satellite DPs is a key feature of graphene 1DSL that are not found in graphene 2DSL or semiconductor 2DEG 1DSL systems. Generally, the scaling factor of v_y of l-th Dirac cone $|f_l| = |v_y/v_F|$ for a given SL modulation u is given by

$$|f_l| = \left| \frac{u}{2\pi l}\left[1 - (-1)^l e^{\frac{iu}{2}}\right]\left(\frac{1}{u - 2\pi l} - \frac{1}{u + 2\pi l}\right)\right| \tag{3}$$

From this formula, by setting $|f_l| = 0$ we can easily show that the flattening of general l-th satellite Dirac cone ($l = 0$ for the main Dirac cone) occurs when

$$u = \begin{cases} 4\pi N, & (\text{even } l, \text{including } l = 0) \\ 4\pi N + 2\pi, & (\text{odd } l) \end{cases}, \tag{4}$$

except $u = 2\pi l$. N is a positive integer. Therefore, a main/satellite Dirac point flattens/unflattens in a periodicity of $\Delta u = 4\pi$. Each time anisotropic band flattening happens, a pair of side Dirac points starts to appear. (Figure 2.6, 2.7)

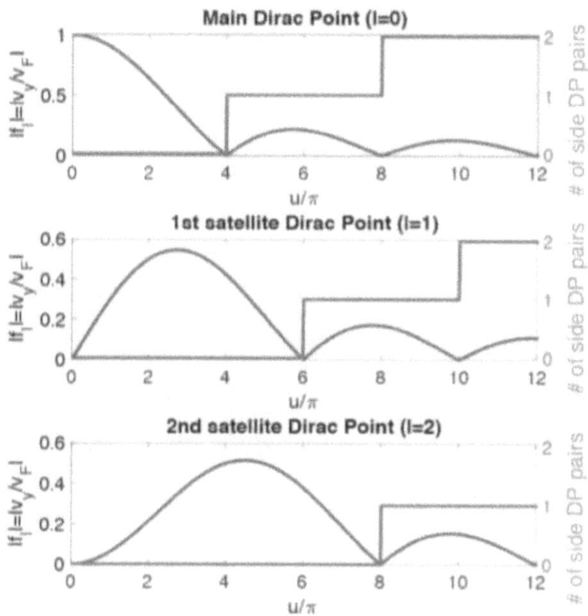

Figure 2.6 Scaling factor of Dirac velocity in the k_y direction of the l-th Dirac cone $|f_l| = |v_y/v_F|$ (Eq. 3) and the number of pairs of its side cones, as a function of u.

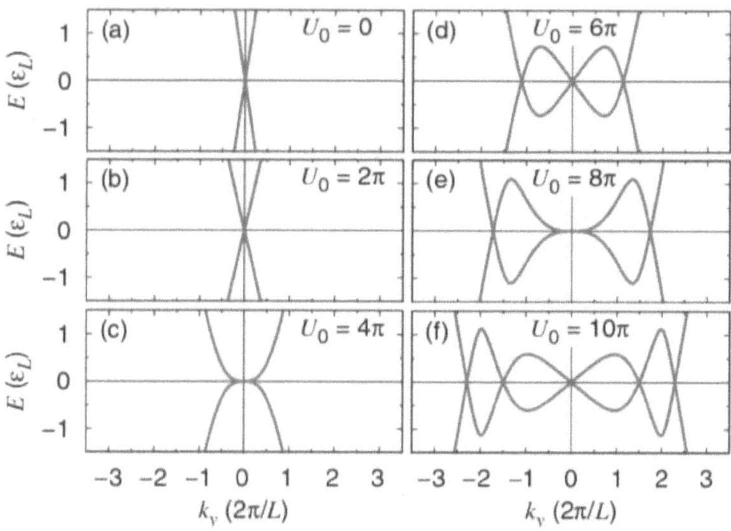

Figure 2.7 Electron energy versus k_y with $k_x = 0$ for various strengths of SL modulation. Reprinted from [4]. Here $U_0 = \frac{V_0 L}{v_F \hbar} = u$ in our notation. ϵ_L is an energy unit defined by $\epsilon_L = \frac{L}{v_F \hbar}$

Finally, it is worth mentioning that for all values of SL modulation u, the band structure of graphene 1DSL exists in an "envelope" that "contains" the satellite Dirac points. As a result, open orbits along the k_x direction exist at high enough energies (Figure 2.4) and have repercussions on the transport properties of the graphene 1DSL system.

2.3.2 Simulated R_{xx} and R_{yy}

Based on the band structures investigated above, we calculated the conductivity in the framework of Boltzmann transport theory with the relaxation time approximation [5]

$$\sigma_{ii} = -\frac{e^2}{2\pi} \sum_n \int \mathbf{k} d\mathbf{k} \left(\frac{\partial f_{n\mathbf{k}}}{\partial \epsilon_{n\mathbf{k}}}\right) \tau_{\mathbf{k}} v_{i,n\mathbf{k}}^2 \tag{5}$$

where $i = (x, y)$, n is the band index, $\mathbf{k} = (k_x, k_y)$ is the wave number, $f_{n\mathbf{k}}$ and $\epsilon_{n\mathbf{k}}$ are Fermi-Dirac distribution and band energies, respectively, $\tau_{\mathbf{k}}$ is the electron relaxation time, and $v_{i,n\mathbf{k}} = \frac{1}{\hbar} \frac{\partial \epsilon_{n\mathbf{k}}}{\partial k_i}$ is the electron velocity of n-th band. The relaxation time approximation introduces a single time scale $\tau_{\mathbf{k}}$ for the relaxation from non-equilibrium to equilibrium distribution, and it is a valid approximation in most of the diffusive regime, such as weakly disordered graphene, where conductance occurs due to elastic scattering. In this calculation, we further assumed an isotropic relaxation time, i.e., $\tau_{\mathbf{k}} = \tau$, since detailed studies have shown that the relaxation time is direction independent to a good approximation, even in the substantially anisotropic systems.[6]

From the calculated conductivity, we plot the resistance ($R_{ii} \propto 1/\sigma_{ii}$) of the graphene 1D SL system as a function of carrier density n and SL modulation strength u in Figure 2.8. We can clearly see the suppression of the resistance R_{xx} by the flattening of the main (labelled by A,C), as well as the satellite-Dirac points (labelled by B) as band flattening results in an increase in the density of states. In contrast, R_{xx} reaches maximum when the DP is unflattened. As we trace along the dashed line corresponding to the main DP, R_{xx} goes from maximum at $u = 2\pi$ (unflattened) to minimum at $u = 4\pi$ (flattened,"A") to another maximum at $u = 6\pi$ (unflattened) to another minimum at $u = 8\pi$ (flattened,"C"). Then we trace along the dashed line corresponding to the 1st satellite DP, R_{xx} goes from maximum at $u = 4\pi$ (unflattened) to minimum at $u = 6\pi$ (flattened,"B"), then to maximum at $u = 8\pi$.In both cases, the maximum/minimum cycle has a $\Delta u = 4\pi$ periodicity, echoing the $\Delta u = 4\pi$ periodicity in the flattening/unflattening of DPs as explained in Eq. 4 and Figure 2.6. The higher order DPs have no conspicuous transport features at low u as the SL modulation is too weak for higher-order satellite DP features to develop.

Compared to the scale-like pattern of R_{xx}, R_{yy} is mostly featureless outside $n = 0$. The lack of transport features corresponding to satellite DPs can be attributed to the presence of open orbits that, as shown in a contour plot of band structure (Figure 2.4), happen at fixed, finite k_y, providing transport channels in the y direction that are unrelated to the flattening/unflattening of satellite DPs. At $n = 0$, R_{yy} shows a maximum at $u = 4\pi$, the SL modulation strength at which R_{xx} shows a minimum. Band flattening in k_y direction (but not in k_x direction) results in a nearly zero carrier group velocity along the k_y direction $v_y \sim \dfrac{\partial E}{\partial k_y} = 0$ and a R_{yy} maxima. However, we don't see another R_{yy} maximum at $u = 8\pi$, despite the main DP being anisotropically flattened as well. This is explained by the presence of 2 side DPs associated with the main DP at $k_x = 0, k_y \neq 0$ at CNP (Figure 2.7d). As these side Dirac cones are not flattened like the main Dirac cone does, we should not expect R_{yy} maxima at $u = 8\pi, 12\pi$ ... as in $u = 4\pi$.

In summary, based on simulations of R_{xx} and R_{yy} we should expect transport data of a graphene 1DSL device to be highly anisotropic and provide evidence for novel band structure features including the cyclic flattening/unflattening of main and 1st satellite Dirac cones, open orbits at finite energies, and side DPs associated with the main DP.

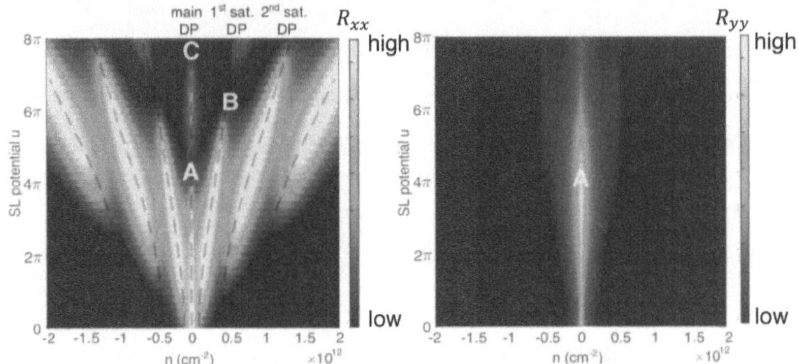

Figure 2.8 Calculated R_{xx} and R_{yy} for a $L = 55nm$ graphene 1DSL device with $W_w = 0.5L$ using relaxation time approximation Dashed Lines indicate R_{xx} features that correspond to the same main/satellite DP.

2.4 The case of $W_w \neq 0.5L$

2.4.1 Evolution of band structure as u and W_w changes

Although $W_w = 0.5L$ is a relatively simple system with particle-hole symmetry, making it the prime target for the first step in a graphene 1DSL experimental study, it is difficult to ensure that the superlattice potential applied on to graphene exactly matches the $W_w = 0.5L$ condition in a real-life device. Therefore, it is important to know how band structures features at $W_w = 0.5L$ would change in cases where $W_w \neq 0.5L$.

Figure 2.9 shows the band structures of graphene 1DSL at $W_w = 0.5L, 0.6L, 0.7L, 0.8L$ at $u = 2\pi, 4\pi, 6\pi$. Figure 2.10 shows the slices of these band structures along $k_x = 0$. For $W_w \neq 0.5L$ we do not expect the conduction and valence bands to be symmetric about each other. Nevertheless, for $u = 2\pi$ the band structure at $W_w = 0.6L, 0.7L, 0.8L$ are almost identical to that of $W_w = 0.5L$ with band structure highly symmetrical about $E = 0$.

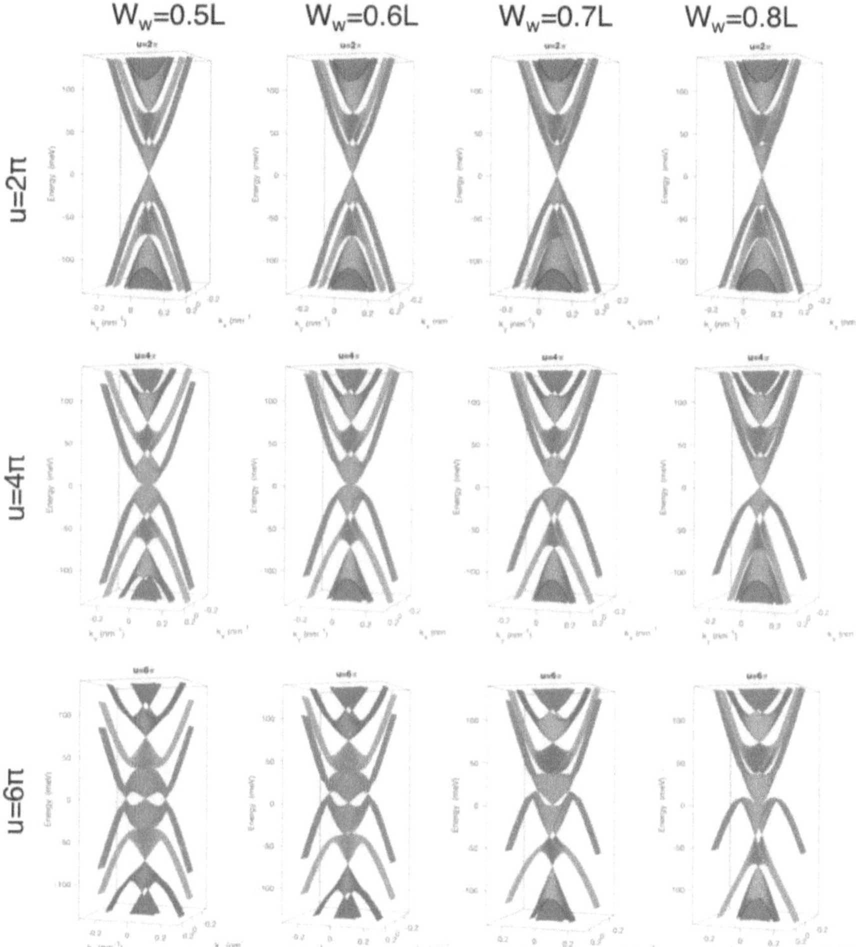

Figure 2.9 Band structure of a graphene 1DSL with $L = 55nm$ at $u = 2\pi, 4\pi, 6\pi$ for $W_w/L = 0.5, 0.6, 0.7, 0.8$.

For $u = 4\pi$ the particle-hole asymmetry is more visible in the band structure at $W_w \neq 0.5L$. At $W_w = 0.6L$, although the main DP is anisotropically flattened as in the case of $W_w = 0.5L$, the 1st valence band is "flatter" than the 1st conduction band at the main DP. At $W_w =$

$0.7L$ the 1st conduction band becomes even less flat, such that at the main DP has discontinuous $\frac{\partial E}{\partial k_y}$ at the CNP. At $W_w = 0.8L$ both the 1st conduction and valence bands have discontinuous $\frac{\partial E}{\partial k_y}$ at the CNP, and the main DP resembles more of a (albeit very distorted) unflattened Dirac cone than the classic anisotropically flattened Dirac cone in $W_w = 0.5L$. It is also worth noticing that for $W_w > 0.5L$ the 3rd valence satellite DP becomes less and less developed as W_w/L increases.

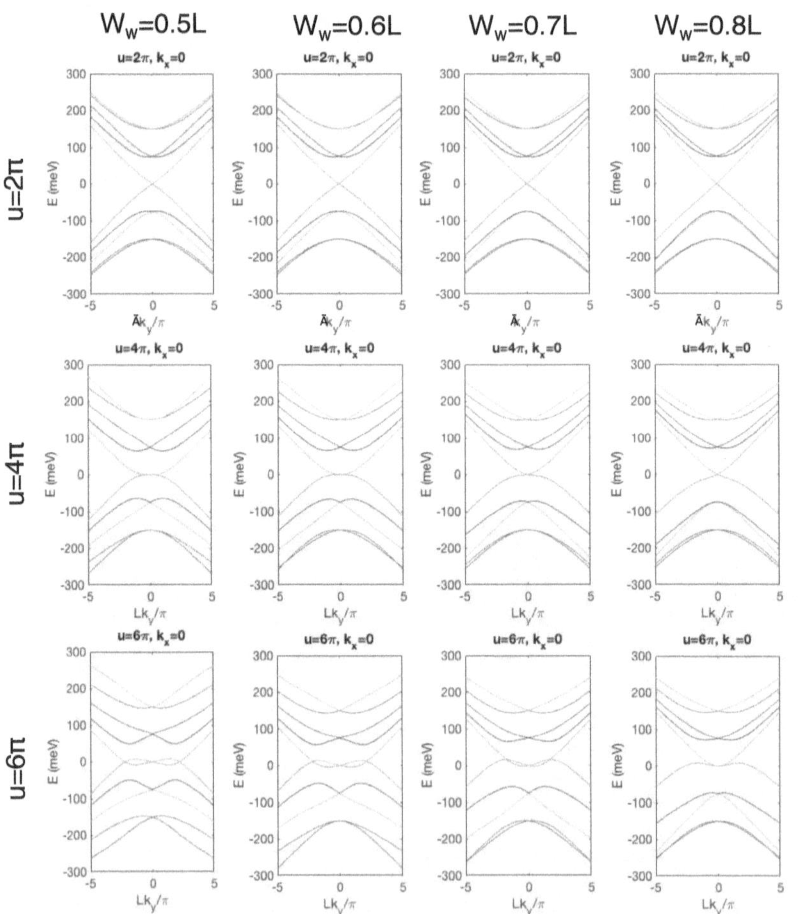

Figure 2.10 Slices of graphene 1DSL band structures at $u = 2\pi, 4\pi, 6\pi$ along $k_x = 0$ for $W_w/L = 0.5, 0.6, 0.7, 0.8$

For $u = 6\pi$, $W_w = 0.6L$ the two side DPs associated with the main DP in the $W_w = 0.5L$ case still exist, but the two side DPs occur at slightly higher energies than the main DP, which still lies at E=0. Moreover, the k_y locations of the side DPs become closer to the main DP than in the $W_w = 0.5L$ case. As W_w/L increases the side DPs and the main DP move closer in k_y. At $W_w = 0.8L$ the side DPs have disappeared and the 1st conduction and 1st valence bands become flat at E=0! As in the $u = 4\pi$ case, we also notice that for $W_w > 0.5L$ the 3rd valence satellite DP (and to some extent, the 2nd valence satellite DP) becomes less and less developed as W_w/L increases.

2.4.2 Simulated R_{xx} and R_{yy}

Based on the band structures above, we further calculate R_{xx} and R_{yy} as a function of carrier density n and SL potential u for cases where $W_w \neq 0.5L$. Electron-hole symmetry is broken in these cases, as seen in the lack of symmetry about the $n = 0$ line in the R_{xx} and R_{yy} maps. At $u = 6\pi$ and $W_w \neq 0.5L$, the R_{xx} maximum at CNP (marked by "a"), does not occur at zero carrier density, and at $W_w = 0.8L$ it is hard to distinguish such a feature at all. This agrees with the shape of the band structure in Figure 2.10 where the 1st mini Dirac points get "flattened" at $u = 6\pi, W_w = 0.8L$.

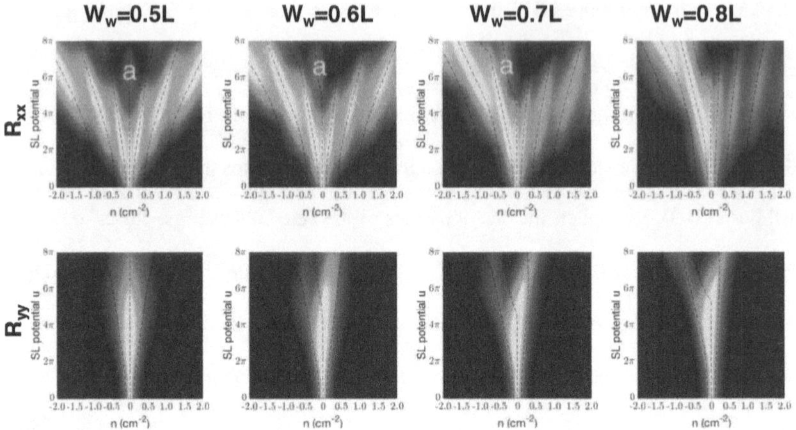

Figure 2.11 Calculated R_{xx} and R_{yy} for a $L = 55nm$ graphene 1DSL device with $W_w = 0.5L, 0.6L, 0.7L, 0.8L$ using relaxation time approximation Dashed Lines indicate R_{xx} features that correspond to the same main/satellite DP. Notice that the CNP does not coincide with zero carrier density for $W_w \neq 0.5L$.

2.5 Summary

Dimensional mismatch between a 1D superlattice and a 2D material (graphene) results in transport anisotropy and heavy dependence of band structure geometry on strength of superlattice modulation. Both are unseen in 2D superlattice on graphene. This could inspire studies of 1DSL and 2DSL hosted on 3D materials.

At equal widths of 1DSL "well" and "barrier", or $W_w = 0.5L$, four types of Dirac points can be identified: main DPs, satellite DPs, side DPs of main DP and side DPs of satellite DP. The former two undergo anisotropic flattening and unflattening cycles in k_y as SL modulation increases with periodicity $\Delta u = 4\pi$. Simulated resistances R_{xx} and R_{yy} are highly anisotropic, and R_{xx} data shows a scale-like pattern related to the cyclic flattening/unflattening of main and satellite DPs.

At unequal widths of 1DSL "well" and "barrier", or $W_w \neq 0.5L$, particle-hole symmetry is broken. The flat bands and side DPs at CNP may evolve into exotic shapes, resulting in asymmetrical R_{xx} and R_{yy} maps.

Chapter 3. Previous studies on 2D systems with anisotropic transport

3.1 Low-symmetry 2D materials

As discussed in Chapter 2, one of the greatest promises of graphene 1DSL is the ability to induce transport anisotropy, a property that occur naturally in a class of 2D materials called "low-symmetry 2D materials".[1] Unlike the common 2D materials, such as graphene, hexagonal boron nitride (hBN) and certain TMDCS like molybdenum disulfide (MoS_2) that have similar electrical, optical and phonon properties along different in-plane crystal directions, low-symmetry 2D materials that include black phosphorous (also called phosphorene in single-layer form) and its alloys with arsenic, as well as monochalcogenides of group IV elements such as SnSe and certain low-symmetry TMDCs like ReS_2, show different electrical, optical, thermal and piezoelectric properties along different in-plane crystal directions with numerous potential applications. .[1] In particular, transport anisotropy has been proposed as a route to bio-inspired electronics, neuromorphic computing [2] and to observing anisotropic quantum Hall effects.

The first low-symmetry 2D material to be widely studied is black phosphorus, which was previously explored as a bulk semiconductor in the 1980s [3] and later rediscovered from the perspective of a 2D material in 2014. [4] Black phosphorous has a puckered orthorhombic crystal structure. Compared to graphene that hosts a six-fold in-plane rotational symmetry (D_{6h}), this symmetry is reduced to two-fold in black phosphorus (D_{2h}). As a result, there is significant transport anisotropy between current travelling in the armchair and zigzag directions, with the carrier conductivity in the armchair direction to be about one order of magnitude higher than in the zigzag direction[5].

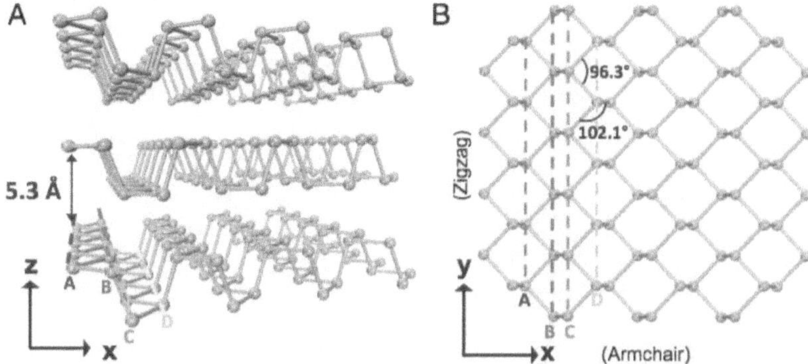

Figure 3.1 Crystal structure of black phosphorus Reprinted from [6]. (A) Side view of the black phosphorus crystal lattice. Interlayer spacing = 0.53nm. (B) Top view of single layer black phosphorus, also known as phosphorene. The armchair and zigzag directions are labelled.

Unlike graphene which is a zero-gap semimetal, few-layer black phosphorus is a direct band gap semiconductor with band gap varying from 0.33eV for 10-layer to 2.0eV for monolayer.[7] Besides number of layers, external electric field[8], doping[9], pressure[10] and uniaxial strain[11] are also able to tune the band gap in black phosphorous and even change it to an indirect band gap semiconductor, semimetal or metal. Few-layer high quality black phosphorus sandwiched between two hBN flakes is able to achieve hole mobility of around 5000cm^{2}V^{-1}s^{-1} at room temperature at about 20000cm^{2}V^{-1}s^{-1} at ~10K[7], making it possible to design field-effect transistors with on/off ratio up to 10^5.[6]

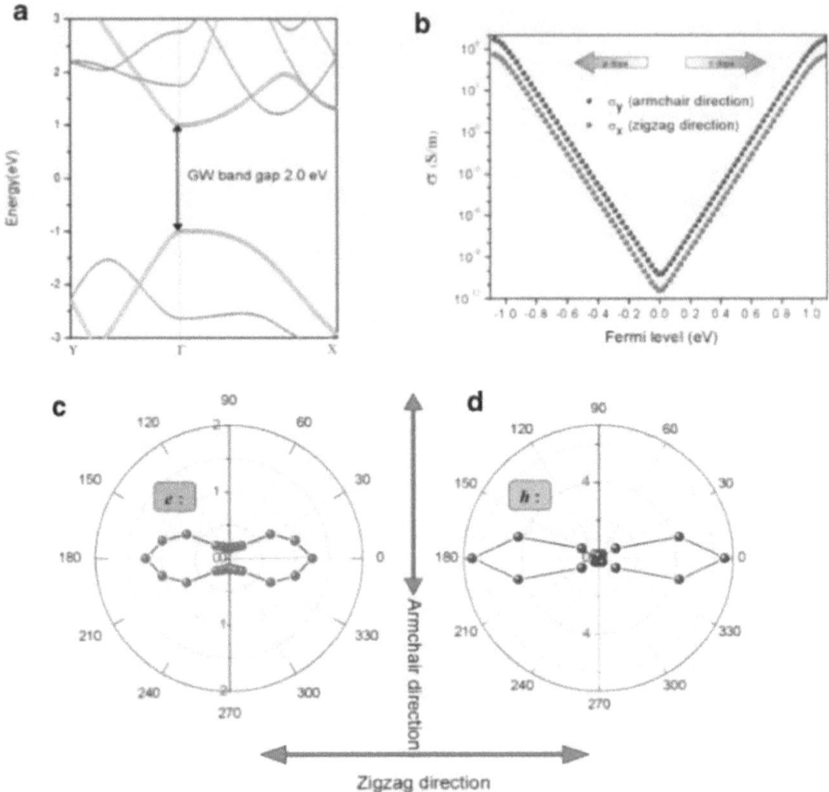

Figure 3.2 Transport anisotropy in black phosphorus Reprinted from [6]. (a) Calculated band structure of suspended monolayer phosphorene, showing a direct band gap. (b) Carrier conductance of monolayer black phosphorus calculated by the Drude model. (c)(d) Effective mass of electrons and holes versus spatial direction. The effective mass along the armchair direction is about one order of magnitude smaller than that along the zigzag direction.

Another group of materials called group IV monochalcogenides also have the puckered orthorhombic crystal structure in black phosphorous, allowing for anisotropic transport between armchair and zigzag directions.[1] Group IV monochalcogenides consist of the compound MX where M =Ge, Sn and X= S, Se. The difference in the M and X atoms means that in Group IV monochalcogenides the inversion symmetry is broken, reducing the point group to C_{2v} and allowing for the occurrence of spontaneous in-plane electric polarization that is predicted to enable 2D ferroelectricity[12].

Group 7 transitional metal dichalcogenides such has ReS_2[13] and $ReSe_2$ [14]have unique crystal structures that are different from the puckered orthorhombic seen in black phosphorus and group IV monochalcogenides, yet also different from their more common Group 6 counterparts such as MoS_2 and WSe_2. ReS_2 and $ReSe_2$ host "Re chains" arranged in parallel, and experiments have shown transport anisotropy between directions along and perpendicular to the "Re chains" in ReS_2 .[15]

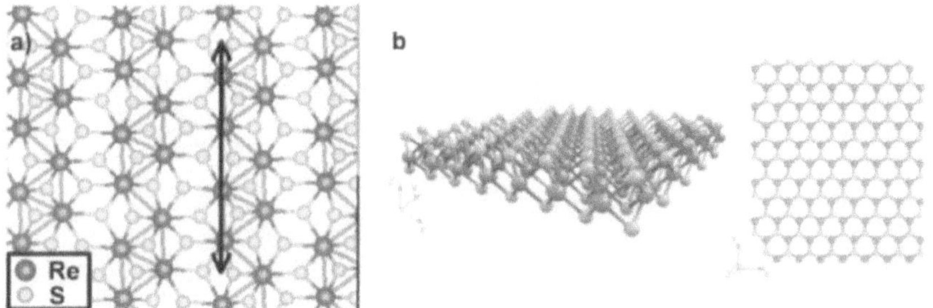

Figure 3.3 Crystal structure of ReS_2 versus MoS_2 (a) Ball-and-stick model of ReS_2 monolayer with direction of the "Re chain" represented by the black double row. Reprinted from[16]. (b) Ball-and-stick model of MoS_2 monolayer.

Despite the wide-ranging promises of these low-symmetry 2D materials, they suffer from numerous practical issues including oxidation susceptibility[17, 18], relatively low carrier mobilities compared to graphene [7] and difficulty in fabricating electric contacts to metal.[19] Therefore, graphene 1DSL demonstrates a viable path towards realizing transport anisotropy with high electron mobility and dynamically tunable anisotropy that could invigorate and motivate new thinking in the quest for low-symmetry 2D materials.

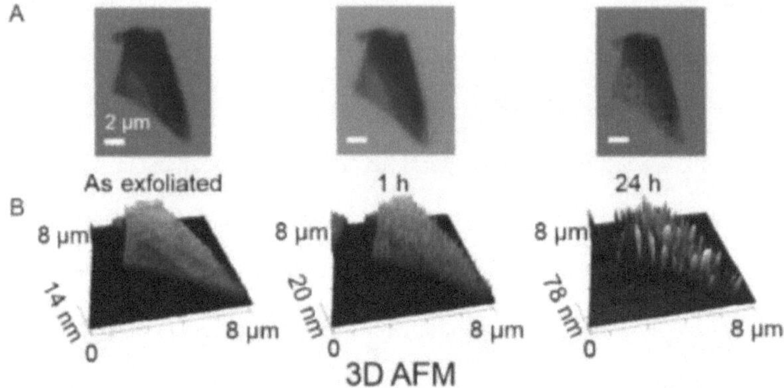

Figure 3.4 Degradation of black phosphorous in air Reprinted from [17](A) Optical microscope image of the same black phosphorous flake immediately after exfoliation, 1h later, and 24h later. Bubble-like features have formed 24h after exfoliation. (B) AFM topography images of the same flake in (A), taken at the same time intervals, showing increased surface roughness as atmospheric exposure time increases.

3.2 Dubey et al's work on Graphene 1DSL system

The first experimental study on graphene 1DSL system was done by Dubey et al in 2013.[20] A graphene Hall bar device is fabricated using the gate patterning scheme as explained in Chapter 1.3 with a silicon back gate and a top gate defined by a sequence of equally spaced palladium lines that are ~27nm wide and 150nm apart. Graphene is sandwiched between a thermally grown SiO_2 bottom dielectric and an Al_2O_3 top dielectric made by atomic layer deposition.

Figure 3.5(b) shows the measured resistance along the wave vector direction of 1DSL (in the language of Chapter 2, R_{xx}), as a function of top and back gate voltages, measured at

T=300mK. Two scale-like resistance patterns symmetrical about $(V_{tg}, V_{bg}) = (0V, 0V)$ can be seen, as predicted by the band structure calculations in Figure 2.8 .In addition, there is a vertical white line indicating the presence of resistance maximum at $V_{bg} = -2V$. Fig. 3.5(c) shows resistance at two constant V_{tg} values as a function of V_{bg}. At $V_{tg} = -0.1V$ where the superlattice modulation is not turned on, the resistance profile looks like that of pristine graphene with one single, narrow peak at CNP. At $V_{tg} = 2.6V$ the resistance shows some oscillation at $V_{bg} < 0$, but the amplitude of the oscillation is very small compared to the background resistance. Fig 3.5 (d) shows the evolution of the resistance oscillation at three different temperatures. The oscillations become largely smoothed out at around 100K that corresponds to 8.3meV, which is on the same order of magnitude as the SL modulation energy scale of $E_{SL} = \frac{\hbar v_F}{d} = 4.4$ meV, suggesting that the oscillations arise from the effects of the 1DSL potential. Dubey et al attributed these resistance oscillations to band structure modifications brought by 1DSL and specifically made a contrast between band structure modifications versus Fabry-Pérot resonance of SL, which is seen in some graphene p-n junction devices [21]. The argument against Fabry-Pérot resonance is as follows: the phase coherence length at 300mK is only 0.6μm, or 4 times the SL periodicity, and would be even shorter at higher temperatures. Since the device measures 40 periods of SL, it is unlikely that resistance measurement would contain any information about phase coherence. The existence of resistance oscillation at 100K, though very weak, is further confirmation that the oscillations cannot come from coherent Fabry-Pérot resonances.

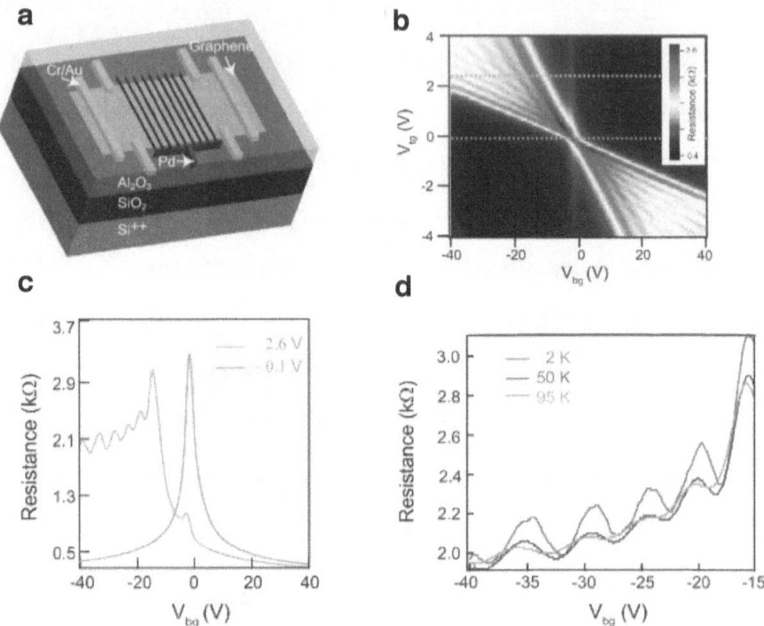

Figure 3.5 Transport measurements from Dubey's device Reprinted from [20]. (a) Device

Schematics. Doped Si is the back gate (V_{bg}) and the Pd lines is the top gate (V_{tg}). (b)

Resistance R_{xx} measured as a function of V_{tg} and V_{bg} at 300mK. (c) Resistance R_{xx}

along the green and blue traces in (b). Notice that along this trace neither carrier density nor

strength of SL modulation u is constant. (d) Line plot of R_{xx} as a function of V_{bg} at three

different temperature with $V_{tg} = 2.6V$ (green line in (b,c))

Having established that the resistance oscillations are caused by band structure modification,

Dubey et al tried to compare their experimental data with the theoretical model of graphene

1DSL developed by Barbier et al[22]. In a gate patterned superlattice device, the strength of

SL modulation depends on the voltages of both gates, and equi-u contours in the (V_{tg}, V_{bg})

space are curved. Figs 3.6(b,c) shows the measured conductance as a function of Fermi

energy at $u = 6\pi$ and $u = 26\pi$ respectively. While the $u = 26\pi$ data shows significant conductance oscillation, there are no conductance features in the $u = 6\pi$ data other than a minimum at CNP, which disagrees with the predictions in Figure 2.8 that suggest the existence of R_{xx} peaks, and thus conductance minima, at the 2nd satellite DP at $u = 6\pi$. This suggests possible error in estimating SL modulation strength u from (V_{tg}, V_{bg}), or limitations to the data quality as seen by the carrier mobility of only 6000 cm²V⁻¹s⁻¹, corresponding to a mean free path of ~70nm which is shorter than the SL period of 150nm. Furthermore, Dubey et al erroneously believed that the resistance oscillation at constant SL modulation versus Fermi energy is a result of the side Dirac points associated to the main Dirac point, instead of the satellite Dirac points. Although side Dirac points do have effects on resistance, these effects are only seen in R_{yy} data where an absence of resistance maxima at $u = 8\pi$ is expected (See Chapter 2.3.2), not in R_{xx} data. By relating the increase in the number of R_{xx} peaks as a function of u directly to the number of side DPs at charge neutrality point, the authors show a superficial understanding of the rich band structure features available in the graphene 1DSL system.

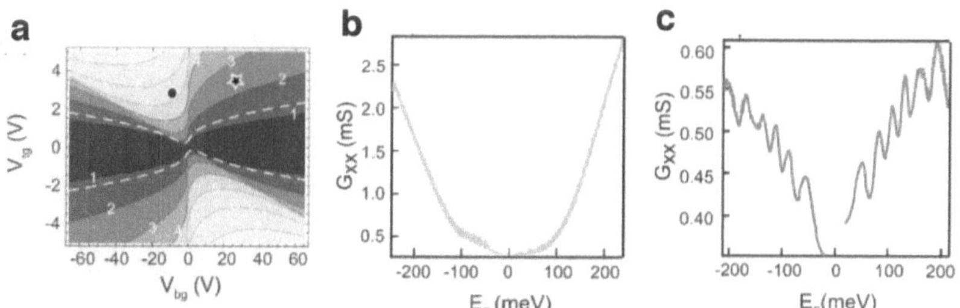

Figure 3.6 Conductance versus Fermi energy at fixed SL modulation strength Reprinted from [20]. (a) Calculated height of potential barrier (u/π) as a function of V_{tg} and V_{bg}. Notice that the equi-u contours are not linear. (b)(c) Measured conductance as a function of Fermi energy at "$u = 6\pi$" and "$u = 26\pi$", respectively, according to the calculations in (a).

3.3 Drienovsky et al's work on Graphene 1DSL system

In 2014, Drienovsky et al did another experimental study on graphene 1DSL. [23]The device architecture is similar to that of Dubey's device, with Al_2O_3 as top dielectric, SiO_2 as bottom dielectric, and a superlattice pitch of 100nm. Unlike Dubey's device with 40 periods of SL, Drienovsky's device has only 4 periods. In addition, large chunks of the transport channel are not affected by superlattice potential.

Figure 3.8 shows the resistance R_{xx} as a function of gate biases. Scale-like patterns of R_{xx} maxima and minima are seen in this device inside the "bipolar" regime (colored purple in Fig 3.7(b)), just as in Dubey's data. However, there are also some resistance features inside the "intermediate" regime (colored green in Fig 3.7(b)) that are not present in Dubey's data. These features are attributed to the ungated regions in the transport channel. Drienovsky et al explained the scale-like resistance pattern not as effects of band structure modification but as Fabry-Pérot resonance, as they calculated transmission function by the recursive Green's function method, in the fully phase coherent limit, at every point in the (V_{TG}, V_{BG}) space, and the resulting pattern is similar to the experimental data. According to Drienovsky et al, electrons will transmit or reflect at the boundaries between the "well" and "barrier" regions of the superlattice. If the phase difference between directly transmitted and twice reflected electron waves is a multiple of 2π, constructive interference happens. The constructive interference condition can be written as

$$2 \int_{well\ or\ barrier} \sqrt{\pi |n(x)|} dx = 2j\pi, j = 1,2 \dots \tag{1}$$

and curves corresponding to such condition in the (V_{TG}, V_{BG}) space are drawn as white dashed lines in Fig 3.8(a), ,and the authors believe that these lines agree with the scale-like experimental data very well.

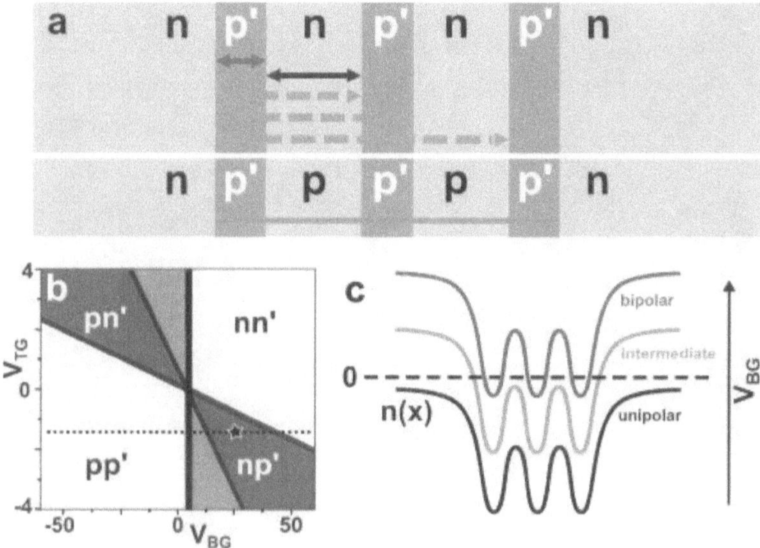

Figure 3.7 Drienovsky's theory of Fabry-Pérot resonance Reprinted from [23]. (a) Schematics of the 1DSL viewed as a series of pn junctions. When the device is in "bipolar regime" as defined in (b), increased reflectivity for ballistic carriers would yield resonances if Eq.(1) is satisfied, as the author argues. (b) Sketch of the predicted resistance regions in the (V_{TG}, V_{BG}) space. The purple region is the "bipolar" regime where neighboring 1DSL "well" and "barriers" have opposite carrier types. Scale-like resistance oscillations happen in this region. The green region is the "intermediate" regime where the carriers inside the 1DSL region have the same type, which is different from the carrier type outside the 1DSL region. This regime is not expected to exist if there are no ungated regions in the channel. The white region is the "unipolar" regime where all carriers in the channel are of the same type. (c) Local carrier density profile $n(x)$ for the three regimes.

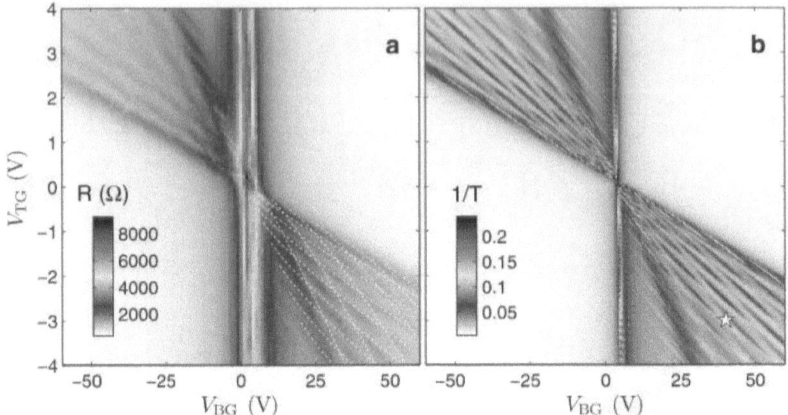

Figure 3.8 Drienovsky's R_{xx} measurement and calculated transmission function
Reprinted from [23]. (a) R_{xx} measured from a device with pitch $L = 100nm$, as a function
of V_{TG} and V_{BG}. White dashed lines indicate the (V_{TG}, V_{BG}) values at which Fabry-Pérot
resonance is expected to happen according to Eq.(1). (b) The inverse of the calculated
transmission function in fully phase coherence limit.

The above analysis based on Fabry-Pérot interference assumes that phase coherence across

the channel, which is not the case in their device, as the carrier mobility is only 7000 cm^{2}V^{-1}s^{-}

^{1}, corresponding to a mean free path of 100nm, which is the length scale of the 1DSL pitch

but not the length scale of the channel length. The authors tried to explain this discrepancy

away by declaring that "the multibarrier system can be understood as independent FP cavities

stringed together, giving rise to a single barrier pattern in the locally ballistic limit", and

therefore "phase coherence over the whole top gated area is not required to reproduce the

general behavior". The authors even used their observation that resistance oscillation persists

at T=100K as evidence that "clearly proves" local ballistic Fabry-Pérot interference

originates in the barriers, while ignoring that the same evidence can be, and has been, used to

argue for the existence of band structure modification. The authors concluded by arguing that "there is no evidence of an artificial band structure and thus a superlattice effect in our multibarrier system".

In 2018 Drienovsky et al repeated this study by encapsulating graphene in boron nitride flakes[24, 25], thus increasing carrier mobility to ~45000 cm^{2}V^{-1}s^{-1}, corresponding to a mean free path of ~800nm, which is longer than the SL periodicity of 100nm but still shorter than the channel length of >5μm. They still attributed resistance oscillation in the transport data with Fabry-Pérot resonance, instead of band structure modification. In addition, they noticed "fine resonance patterns" within the rhombic mesh of Fabry-Pérot oscillations, and they attributed them to "ballistic transport across several potential barriers." They further investigated the oscillation of R_{xx} as a function of perpendicular magnetic field at a given carrier density and SL modulation strength. R_{xx} minima appear when the semiclassical diameters of the carriers $2r_c$ become an integer (up to a 1/4 difference) multiple of the 1DSL periodicity L:

$$2r_c = \left(\lambda - \frac{1}{4}\right) L, \tag{2}$$

with λ being an integer. This oscillation is called commensurability oscillation, or Weiss oscillation, as Weiss first discovered it on a 2DEG-1DSL system[26]. In the magnetotransport data they are able to identify commensurability oscillation minima up to $\lambda = 6$. Although commensurability oscillation exists alongside, and is dwarfed by, Shubnikov-de Haas (sdH) oscillations and quantum Hall oscillations, to the point that at $\lambda = 1$ R_{xx} can even be a local maximum,, the authors are still able to extract commensurability oscillation from the sdH-QHE envelope and observe that it persists to 150K. [25]It is interesting to note that the authors acknowledge the formation of minibands in graphene Landau levels by the superlattice modulation in the quantum explanation of commensurability oscillation (to be explained in Chapter 4), but refuse to acknowledge the existence of minibands, or any band structure modification, at zero magnetic field.

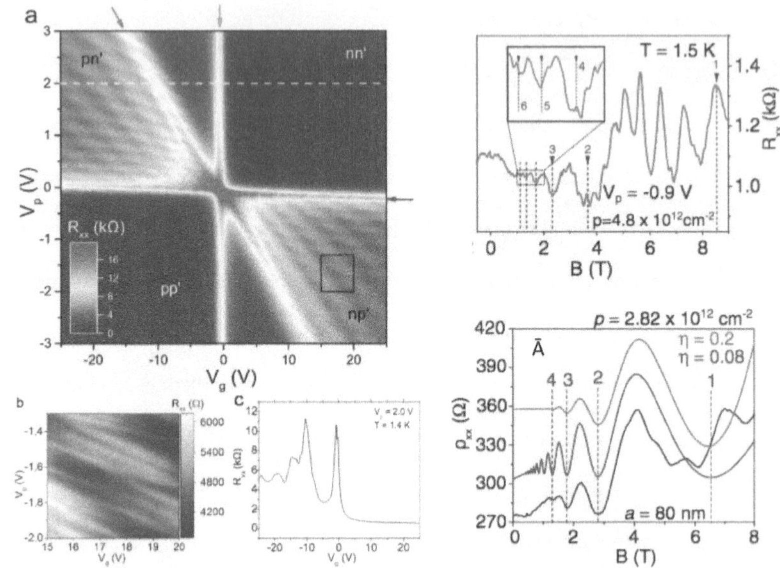

Figure 3.9 Drienovsky's further work on graphene 1DSL Reprinted from [24, 25]. (a) R_{xx} measured from a hBN-encapsulated graphene 1DSL device with pitch $L = 100nm$. (b) R_{xx} inside the black box in (a), showing the "fine resonance patterns" within the rhombic mesh of Fabry-Pérot oscillations. (c) R_{xx} along the white dashed line at $V_p = 2.0V$ (d) R_{xx} versus vertical magnetic field B at fixed SL modulation, carrier density and temperature in a L=80nm device. Expected R_{xx} minima as predicted by Eq. (2) are labelled by blue dashed lines with λ. Commensurability oscillation is dwarfed by sdH and quantum Hall oscillations. (e) Extraction of commensurability oscillation (blue) from raw R_{xx} data (black) after subtracting sdH oscillation (red).

It should be noted that a later work in 2020 by Kang et al, from the same group as Drienovsky, acknowledged the existence of minibands in graphene 1DSL systems.[27]

3.4 Summary

Anisotropy is 2D materials can be found in a class of materials called "low-symmetry 2D materials" that include black phosphorous and its arsenic alloy, monochalcogenides of group IV elements and certain group 7 TMDCs. They have exotic anisotropic electrical, optical, thermal and piezoelectric properties, but also suffer from practical issues such as high oxidation susceptibility compared to graphene.

Previous work on graphene 1DSL has not investigated transport anisotropy, which contains important information on band structure not available from R_{xx}. In addition, there used to be an unsettled debate on the nature of R_{xx} oscillations as a function of gate voltages in graphene 1DSL devices. Dubey et al attribute them to band structure modification, and Drienovsky et al attribute them to Fabry-Pérot resonance. Dubey has a relatively more solid argument for band structure modification, but she misattributes R_{xx} features to the number of side Dirac cones at CNP. Drienovsky's work is notable for his magnetotransport measurements that revealed commensurability oscillations, but his arguments for Fabry-Pérot resonance and against band structure modification is dubious. Therefore, a more detailed study on graphene 1DSL is warranted.

Chapter 4. Electrical measurements of graphene 1D superlattice devices

4.1 Device fabrication

4.1.1 Device Schematics

An improved graphene 1DSL device needs to have high carrier mobilities and ability to conduct transport measurements in both R_{xx} and R_{yy}. Therefore, I design new graphene 1DSL devices based on the schematics of Figure 4.1. The device is based on 2D materials except for the superlattice and the back gate, and is made into an L-shaped double Hall bar, allowing measurements on transport anisotropy.

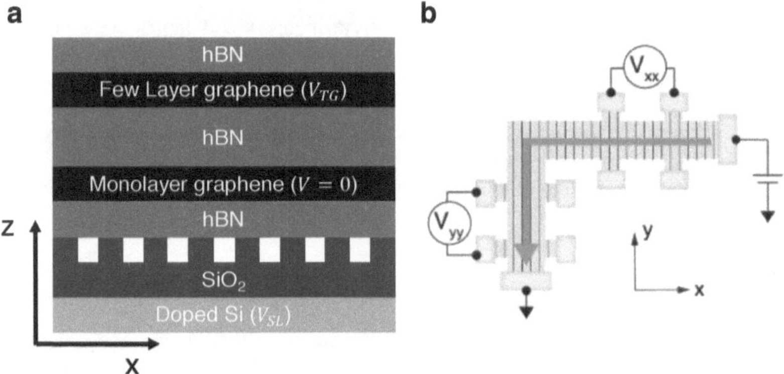

Figure 4.1 Device schematics of the new graphene 1DSL devices Not to scale. (a) View of the device along the x-z plane. (b) View of the device along the x-y plane, showing the L-shaped Hall bar. Green region: graphene channel. Yellow region: graphene to metal contacts. Red arrow: Direction of current flow. Black lines: Lines of 1DSL. $R_{xx} = V_{xx}/I$ and $R_{yy} = V_{yy}/I$ where I is the current in the channel.

4.1.2 A step-by-step guide to fabrication

In our devices, the van der Waals heterostructure and the patterned dielectric superlattice (PDSL) are first fabricated separately before being brought together.

The vdW heterostructure consists of hBN-Few layer graphene (FLG) -hBN-monolayer graphene (MLG) -hBN. Dry transfer techniques using a glass slide containing a PDMS stamp covered by a polypropylene carbonate (PPC) film are used to assemble the stack. Before assembling the stack, the hBN/FLG/MLG flakes are exfoliated on SiO_2/Si chips and checked under optical microscope to ensure that they are free from major defects such as step edges, cracks and wrinkles. The PPC film is brought to contact with the region of SiO_2 chip containing the desired flake on a transfer station at T=40°C~50°C and the PPC slide is then raised up such that the desired flake sticks to the PPC. The top hBN, though not participating in gate biasing nor carrier transport, is necessary as the top layer of the stack since it is much easier to pick up hBN than FLG using PPC. Once the top hBN is picked up the other layers can be easily picked up due to van der Waals attraction. The FLG acts as the top gate of the device. The mid hBN with thickness 30~50nm acts as the top dielectric. The bottom hBN flake, acting as bottom dielectric along with SiO_2, is as thin as ~5nm to maximize the amount of SL modulation felt by the MLG. After picking up all the 5 layers (Fig 4.2 Left), the stack is dropped onto an empty SiO_2/Si chip. During this dropping process, bubbles formed between the individual vdW layers are squeezed out of the stack at a temperature around 80°C. At 120°C the PPC slide is slowly raised up, leaving PPC and the stack on the SiO_2/Si chip. Vacuum annealing of the chip for 20 min at 360°C is used to remove PPC, leaving only the desired stack on the chip (Fig 4.2 Center)

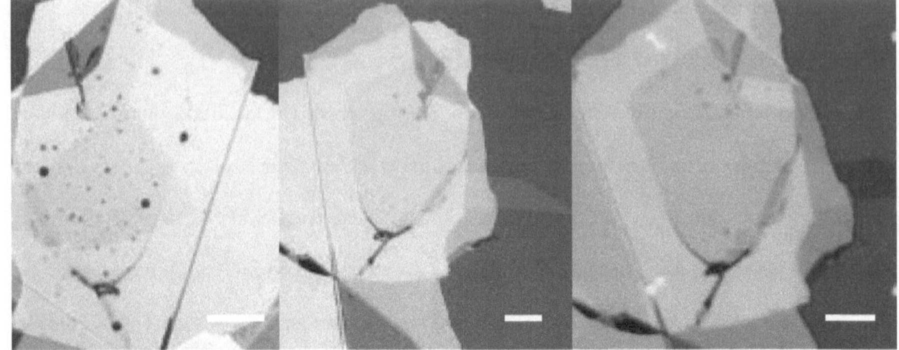

Figure 4.2 Assembled van der Waals heterostructure Left: Optical microscope image of the vdW heterostructure assembled for making the L=47nm device. The stack is attached to the PPC slide. Many bubbles that are trapped between individual vdW layers can be seen. Center: Optical microscope image of the same stack after being dropped onto a SiO_2/Si chip with most of the bubbles pushed out. Right: Optical microscope image of the same stack after being picked up again and dropped onto a SiO_2/Si chip with 1D PDSL. The green square in the middle of the stack is the 10μm × 10μm region with 1D PDSL lines inside. Scale bar = 10μm.

The fabrication of 1D PDSL on a SiO_2/Si substrate is based on the 2D PDSL fabrication techniques developed by Carlos Forsythe[1, 2]. PMMA 495 A2 is spun onto a SiO_2/Si chip at 3500rpm and baked at 180°C for 2 minutes, forming a PMMA layer about 50nm thick. A sequence of straight lines separated by 47nm or 55nm are written using the e-beam lithography system NanoBeam (Nanobeam Ltd. Cambridge, UK) at a dose of ~340 C/m^2 and a beam current at 0.3nA. All these lines are confined in a 10μm × 10μm region. After e-beam writing the chip undergoes sonicated development in IPA:MIBK (3:1) at 3°C for 30 seconds. A very short (~10 seconds, depending on power) O_2 plasma helps etch undeveloped PMMA at the line sites. Then the chip is etched with CHF_3 (40 sccm) + Ar (5 sccm) with 60W forward power and 0W ICP power for 12 cycles. Each cycle consists of a 30 second etch and a 30 second cooling period. This results in an etch of roughly 40nm deep. Afterwards, 3 minutes of O_2 plasma is used to remove the PMMA that remains on the chip. Finally, the chip is cleaned in 160°C Piranha solution (sulfuric acid: 30% H_2O_2 = 3:1 in volume) for 15 min, followed by a 3-minute rinse in DI water and N_2 blow dry.

Using an optimal dose for e-beam lithography is critical to the success of PDSL patterning. Generally, the width of the etched straight lines increases as the e-beam dose increases. But a dose too high can introduce unwanted breaks in the etched lines. Because it is difficult to predict the correct dose beforehand, at each PDSL writing an array of 20 regions named Dose 1, Dose 2…. Dose 20 that are sized 10μm × 10μm, each written with the same line pattern at different doses arranged in a geometric sequence, with $(Dose\ n) = (Dose\ 1) * 1.05^{n-1}$. Furthermore, this array is written at two locations on the 1cm-by-1cm SiO_2/Si chips. Assuming the consistency of the e-beam writer, the two regions that have the same dose should be identical, so I can dice the chip in two halves and use one half for scanning electron microscope (SEM) imaging to find out the optimal dose to use. The corresponding dose on the other half is used in the actual device. As seen in the SEM image insets of Figure 4.3, even in the overdosed situation the width of the etched region is still less than one half of the SL pitch L. Since I want to strive for the case where $W_w/L = 0.5$, the optimal dose to use is the largest dose at which there are no unwanted breaks in etched lines. In both of my $L = 47nm$ and $L = 55nm$ devices the etched region width is only 13nm.

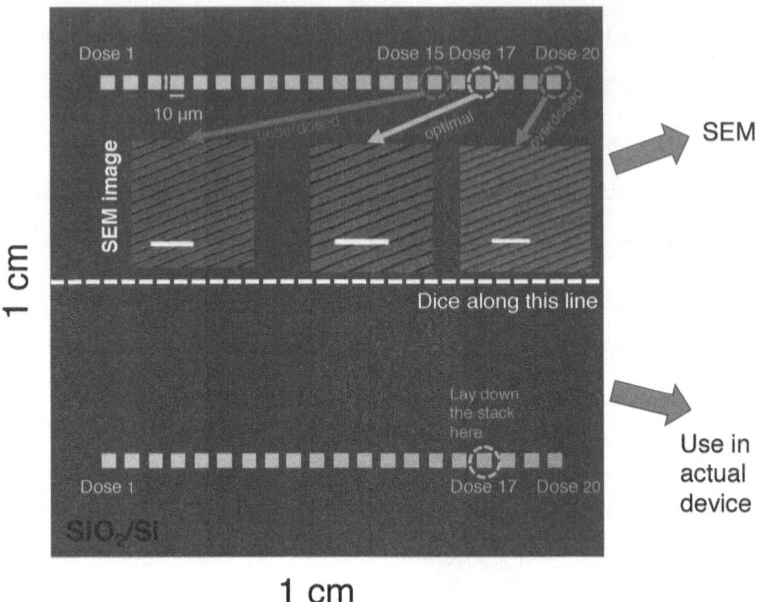

Figure 4.3 Dose testing This is a schematics diagram (not to scale) showing the process of dose testing. Electron dose increases from Dose 1 to Dose 20, each "dose" is a 10µm × 10µm square inside which parallel lines are drawn by the e-beam writer. Inset shows SEM images of fabricated 1D PDSL at three different doses. At Dose 17 the etched black lines are the widest possible without causing any unwanted breaks as in Dose 20. Therefore, Dose 17 on the other half of the chip is used in the actual device. Scale bar=200nm.

With both the vdW heterostructure and the 1D PDSL chip prepared, we use a PPC/PDMS slide to pick up the vdW heterostructure and drop it on top of the 10µm × 10µm region patterned with 1D SL lines. Standard nanolithography techniques are used to define the top gate, shape the Hall bars and fabricate metal-to-graphene edge contacts[3].

4.2 Determination of carrier density n, V_{TG} and V_{SL} offsets, and carrier mobility μ

During a graphene 1DSL device transport measurement, the graphene sheet is kept grounded while the FLG top gate and the doped Si back gate, separated from graphene sheet by dielectric materials, are subject to voltage biases $V_{TG,raw}$ and $V_{SL,raw}$ respectively. Using a simple parallel plate capacitor model, we calculate the average carrier density according to $n = (C_{TG}/e)(V_{TG,raw} - V_{TG,0}) + (C_{SL}/e)(V_{SL,raw} - V_{SL,0})$. The geometric capacitances C_{TG} and C_{SL} are determined by from zero field transport (Fig. 4.4a) and Hall response in R_{yy} (Fig. 4.4b), where a Landau fan emanates from the CNP, enabling us to convert from $V_{TG,raw}$ to n based on the fact that in graphene subject to magnetic field, R_{yy} reaches minimum at filling factors $\nu = \pm 2, \pm 6, \pm 10, \pm 14$ .... The gate voltage offsets $V_{TG,0}$ and $V_{SL,0}$ are determined by requiring that zero field transport should be symmetric about $n = 0$ and $V_{SL} = 0$. In both the $L = 47nm$ and $L = 55nm$ devices $V_{TG,0}$ for the R_{xx} and R_{yy} data differ by 1~2V. The V_{SL} used in this paper is $V_{SL,raw} - V_{SL,0}$.

The strength of carrier density modulation in the channel due to the existence of 1D SL is mostly determined by the SL gate bias V_{SL}. Therefore, in a dual-gated PDSL device we achieve independent tuning of two variables, namely average carrier density n and strength of carrier density modulation Δn, using two experimentally controllable parameters, V_{TG} and V_{SL}.

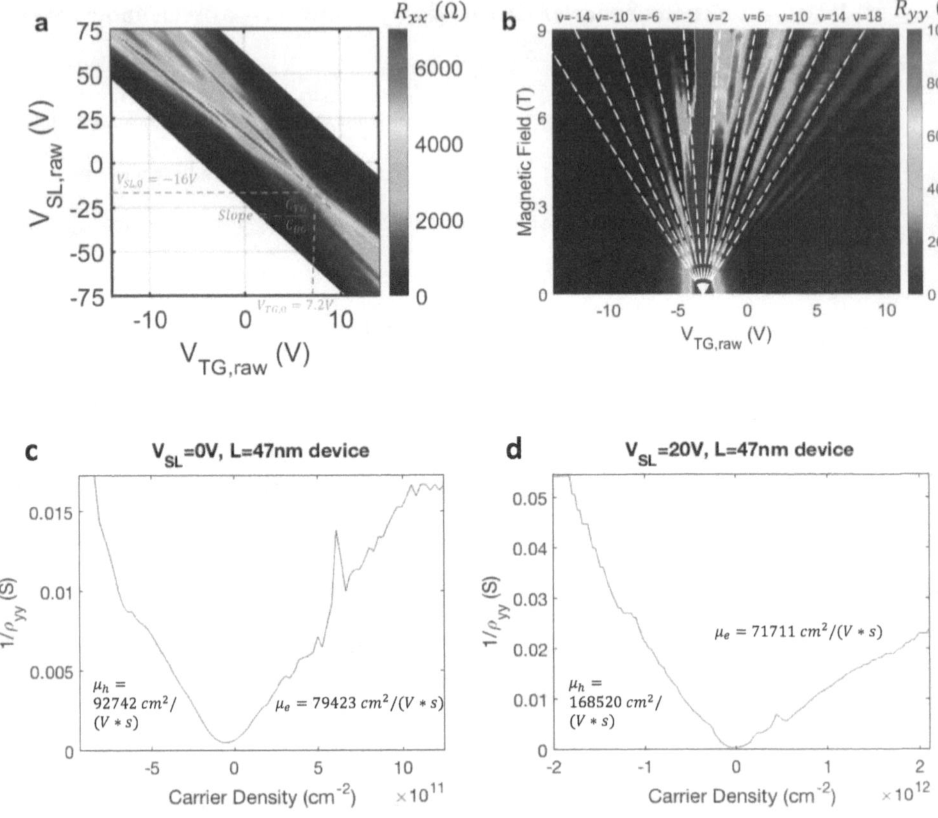

Figure 4.4 Determination of carrier density n, V_{TG} and V_{SL} offsets, and carrier mobility μ (a) R_{xx} as a function of $V_{TG,raw}$ and $V_{SL,raw}$ at $B = 0T$ for the $L = 55nm$ device. The gate voltage offsets $V_{TG,0}$ and $V_{SL,0}$ are located at the "origin" where $n = 0$ and the SL modulation is zero. The ratio of capacitances can be found using the slope of the R_{xx} maximum near the "origin", as this R_{xx} maximum should lie on $n = 0$. (b) R_{yy} as a function of $V_{TG,raw}$ and magnetic field at $V_{SL,raw} = 25V$ ($V_{SL} = 41V$) for the $L = 55nm$ device. White dashed lines trace the Landau fan originating from CNP, enabling us to relate gate voltages to carrier density n. (c,d) Conductance vs. carrier density plots for the $L = $

$47nm$ devices at two different back gate voltages. Electron/hole mobilities calculated from the slope of these curves are annotated next to them.

Given that R_{yy} data is largely featureless aside from the CNP, the carrier mobilities in our devices are estimated using R_{yy} data. Fig. 4.4c shows the conductance vs. carrier density plot for the $L = 47nm$ device at $V_{SL} = 0V$, showing a linear relationship at $|n| < 5 \times 10^{11} cm^{-2}$. The electron and hole carrier mobilities can be extracted from the slope of the conductance vs n curve, according to the equation $\sigma = ne\mu$, and are much larger than the carrier mobilities in previously studied 1DSL devices.

4.3 R_{xx} and R_{yy} at $B = 0T$

4.3.1 Low temperature measurements

Two graphene 1DSL devices with pitches $L = 55nm$ and $L = 47nm$ are measured at $T = 1.7K$ and $T = 2K$ respectively. While the $L = 47nm$ device has all metal-to-graphene contact working, the $L = 55nm$ device has multiple contacts that failed. As a result, the R_{yy} for the $L = 55nm$ device cannot be measured in the usual way, and the measured R_{yy} would include some unknown resistance offset.

a   b

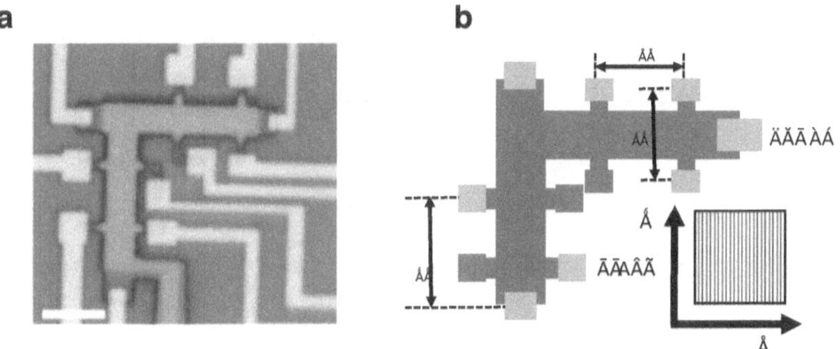

Figure 4.5 Image of a finished device and contact issues (a) Optical microscope image of the $L = 47nm$ device.   : metal contracts and leads. Scale bar = 5μm .Green: Regions

with hBN-FLG-hBN-MLG-hBN. Blue: Regions with hBN-MLG-hBN with the topmost two layers etched away. Electrical contacts to graphene can only be laid on the blue region to avoid any leakage to the top gate (b) Schematics of the $L = 55nm$ device showing the locations of good (green) and bad (red) contacts and the method of measuring $R_{xx} = V_{xx}/I$, $R_{xy} = V_{xy}/I$, and $R_{yy} = V_{yy}/I$. Because the drain is located in the middle of the transport channel, the measured R_{yy} would include an unknown amount of offset. The inset shows the definitions of x and y directions in 1DSL.

Figure 4.6a,b shows measured R_{xx} and R_{yy} from the $L = 55nm$ device, respectively. Figure 4.6c,d shows the calculated R_{xx} and R_{yy} from the $L = 55nm$ device as seen in Figure 2.8 .Overall, the features between experiment and calculation agree very well with each other. In particular, R_{xx} data is highly symmetric about $n = 0$, suggesting that despite the width of the etched SL lines being less than $L/2$, the actual SL potential felt by graphene in the device is very close to $W_w/L = 0.5$. In Figure 4.6a,c, dashed curves trace along R_{xx} features related to the same mains/satellite Dirac point. Along the curves corresponding to the main and 1st satellite Dirac points, R_{xx} undergoes cycles of maxima and minima, with the R_{xx} minima achieved when the main/satellite Dirac point is anisotropically flattened. (Figure 4.6h,i). R_{yy} shows a maximum at $n = 0$, $V_{sl} \sim 40V$, at which shows R_{xx} a minium. The existence of this R_{yy} maximum is due to the flattening of main CNP at $u \sim 4\pi$ in the k_y direction. The absence of R_{yy} maximum at $u \sim 4\pi$ is due to side DPs at CNP. The absence of R_{yy} features outside n=0 is due to the presence of open orbits. More detailed explanation on the relationship between band structure and resistance can be found in Section 2.3.2.

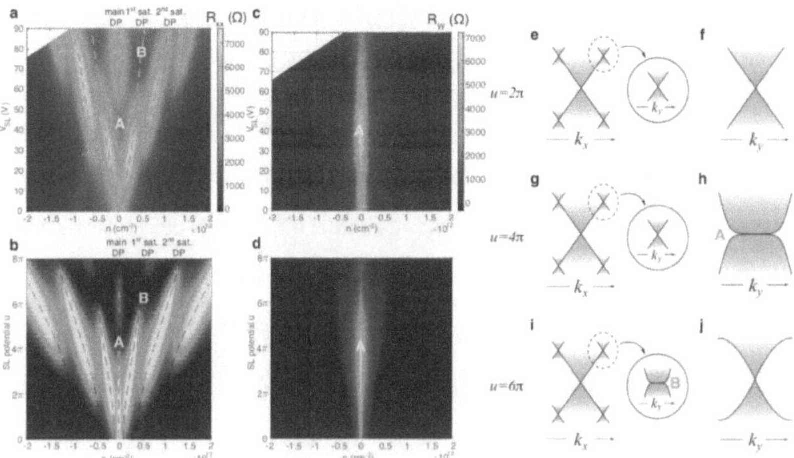

Figure 4.6 Comparison between measured vs calculated R_{xx}, R_{yy} (a) Device resistance R_{xx} as a function of carrier density n and back gate voltage V_{SL}, measured from an $L = 55nm$ device (b) Calculated R_{xx} for $L = 55nm$ device plotted as a function of dimensionless SL strength u and carrier density n. In (a,b) gray dashed lines trace along features related to the same main/satellite Dirac point. (c) R_{yy} measured from the same $L = 55nm$ device. (d) Calculated R_{yy} for $L = 55nm$ device (e~i) Band structures at $u = 2\pi, 4\pi, 6\pi$ sliced along $k_y = 0$ (e,g,i), $k_x = 0$ (f,h,j), and $k_x = \pm\pi/L$ (insets). Throughout this figure, select R_{xx} and R_{yy} extrema are related to the anisotropically flattened Dirac cones by letters A and B.

Figure 4.7 compares data taken from the $L = 47nm$ (a,b) and the $L = 55nm$ (c,d) devices. The R_{xx} data from the $L = 47nm$ device looks similar to the R_{xx} data from the L=55nm device except that the dimensionless strength of SL modulation, $u = V_0 L/\hbar v_F$, is only about 5π at the maximum $V_{SL} = 100V$, while in the $L = 55nm$ device $u \approx 7\pi$ at maximum $V_{SL} = 75V$. This is a natural consequence of the fact that for the same strength of SL modulation V_0 measured in meV, the shorter the SL line spacing L, the smaller the

dimensionless strength of SL modulation, $u = V_0 L/\hbar v_F$. Therefore, in the $L = 47nm$ R_{xx}

data we don't have enough SL modulation to see the flattening of the 1st satellite Dirac cone

($u = 6\pi$, labelled by B,C in Fig 4.7c) as well as the un-flattening of the main cone ($u = 6\pi$,

labelled by δ in Fig 4.7c). The R_{yy} data from the $L = 47nm$ device (Fig. 4.7a) feature a

pair of maxima at $u \approx \pm 4\pi$. However, the R_{yy} data from the L=55nm device shows only a

maximum at $u \approx 4\pi$ ($V_{SL} \approx 40V$) but not at $u \approx -4\pi$ ($V_{SL} \approx -40V$). Although it is

tempting to explain the lack of R_{yy} data symmetry about $V_{SL} = 0V$ by claiming that the

R_{yy} for $L = 55nm$ is measured in the wrong configuration and is not the "correct" R_{yy}, it

should be noted that even for the $L = 47nm$ device R_{yy} at $V_{SL} \approx 40V$ is higher than at

$V_{SL} \approx -40V$. In the graphene 1DSL Hamilton, a $V_{SL} \rightarrow -V_{SL}$ transformation corresponds to

a swap between the "well" and "barrier" regions in the superlattice. [4] Therefore, it is likely

that V(x), the 1DSL potential that the graphene sheet actually feels in the device, is not a

centrosymmetric function about $x = 0$, $V_0/2$ (Figure 2.1), where we define $x = 0$ as the

location of a well-barrier boundary.

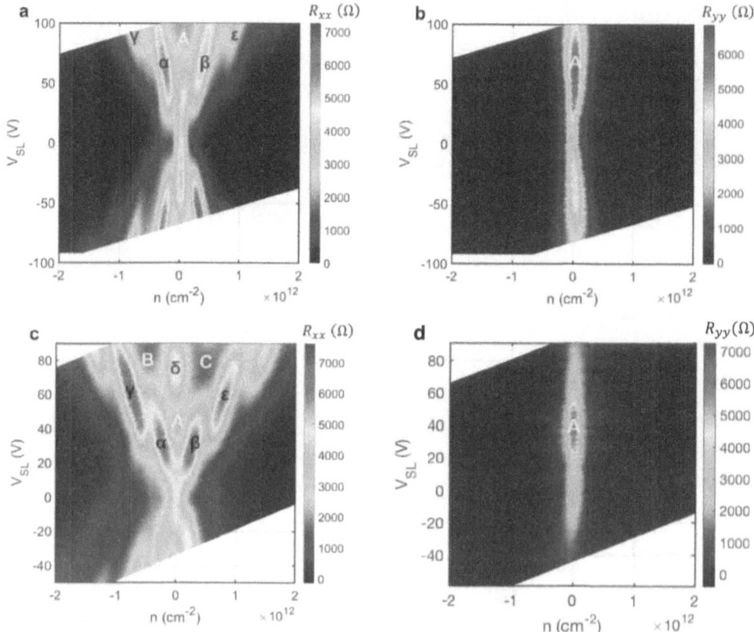

Figure 4.7 Comparison between L=47nm and L=55nm data. (a) R_{xx} for $L = 47nm$ (b) R_{yy} for $L = 47nm$ (c) R_{xx} for $L = 55nm$ (d) R_{yy} for $L = 55nm$. Throughout this figure R_{xx} minima (maxima) that correspond to the same band structure feature are indicated by Latin (Greek) letters, respectively.

4.3.2 Temperature dependence measurements

Figure 4.8 shows the R_{xx} data at T=2K, 20K and 130K for the $L = 55nm$. Very little difference is seen between T=2K and T=20K. At T=130K, an "X" shaped region of R_{xx} maxima is seen. In fact, remnants of the "X" shaped region still show as bumps in R_{xx} at T=235K. This behavior is in stark contrast to the case of graphene 2DSL, where most of the side resistance peaks associated with the superlattice has been smeared out at T=95.5K[1]

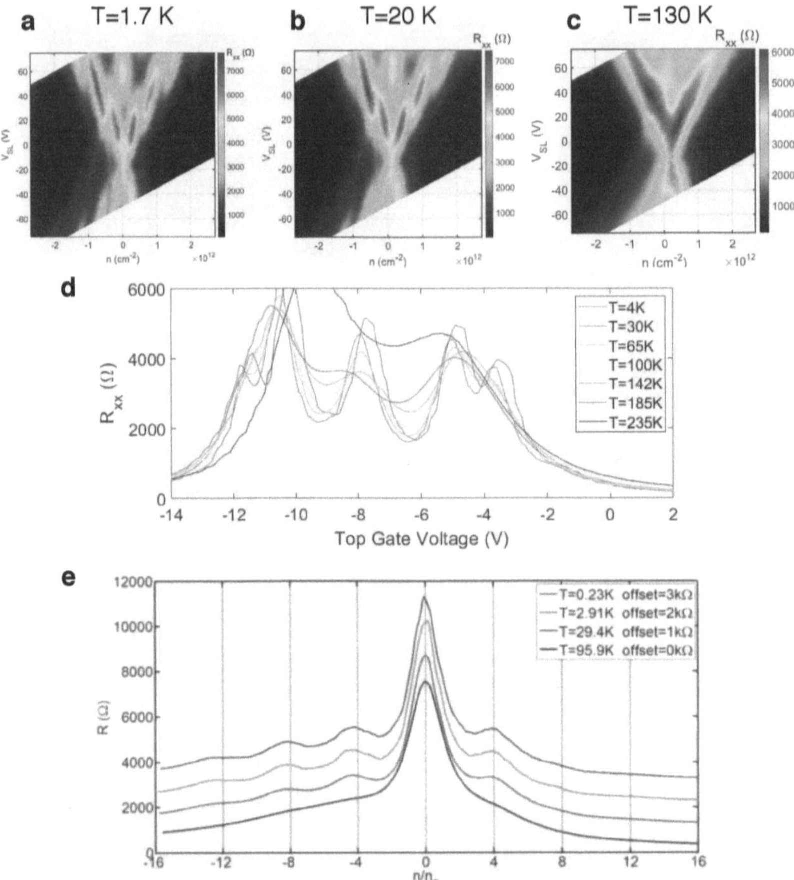

Figure 4.8 Temperature dependence of R_{xx} in graphene 1DSL and 2DSL (a~c) R_{xx} in the $L = 55nm$ graphene 1DSL device at T=1.7K, 20K and 130K, respectively. (d) R_{xx} in the same device with fixed $V_{SL} = 76V$ at several different temperatures (e) Temperature dependence of superlattice induced satellite peaks in a graphene 2DSL device for V_{SL} . Lines plotted with various offsets for clarity. Reprinted from [1]

In Drienovsky et al's interpretation of graphene 1DSL as a sequence of pn junctions, the "X" shaped region corresponds to the transition between "bipolar" and "unipolar" regimes [5]

(Figure 3.7c and 4.9c), and they did measure a similar persistence of R_{xx} maximum at high temperature. In contrast, the region "δ" in Figure 4.7c, deep inside the "bipolar" regime, is mostly smeared out at 130K. Similarly, in Drienovsky's data the resistance peak at the transition between "unipolar" and "polar" regimes are stable against the temperature up to 100K and was explained as "single barrier resonance". [5]The oscillation features inside the "bipolar" regimes are smeared out at 100K.

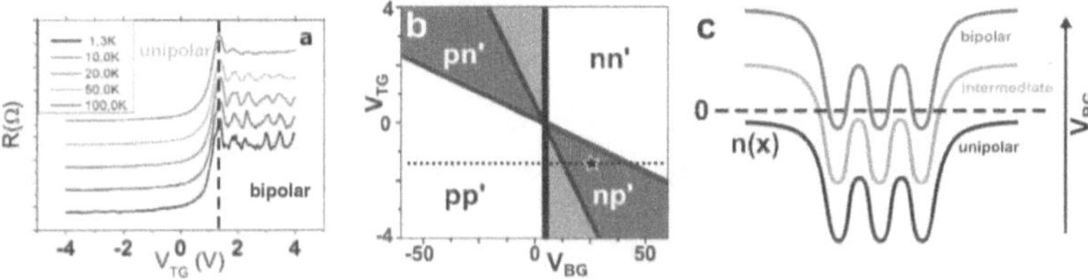

Figure 4.9 Drienovsky's temperature dependence measurements and the pn junction interpretation Reprinted from [5]. (a) Resistance vs top gate voltage in Drienovsky's graphene 1DSL device at various temperatures . (b)(c) Explanation of the terms "bipolar" and "unipolar". In the unipolar regime all carriers in the channel are of the same type, while in the bipolar regime the 1DSL can be seen as a sequence of pn junctions. Same as Figure 3.7(b)(c)

Figure 4.10 (b) shows the R_{yy} measured along the CNP versus V_{SL} at various temperatures. As temperature increases, the peak values of R_{yy} decreases and tend to move towards $V_{SL} = 0$. The $V_{SL} < 0$ peak collapses at around T=4K while the $V_{SL} > 0$ peak still survives at T=100K. Therefore, the absence of R_{yy} maximum in the L=55nm device could be explained by the $V_{SL} < 0$ peak having a "flattening temperature" less than 2K, the temperature at which the measurement took place. A measurement at an even lower temperature could have revealed the existence of the $V_{SL} < 0$ peak.

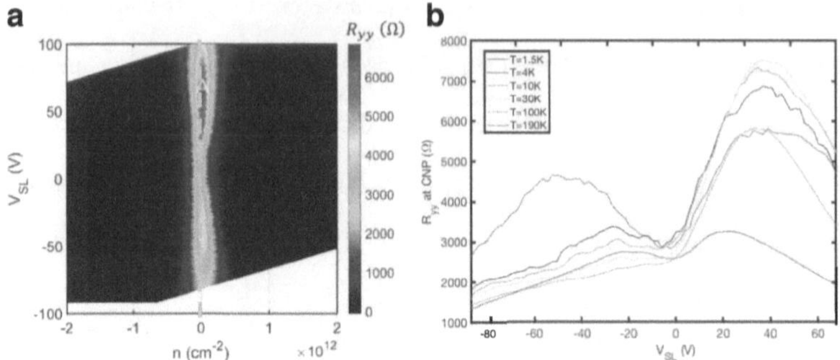

Figure 4.10 Temperature dependence of R_{yy} at CNP (a) R_{yy} for the $L = 47nm$ device, same as Figure 4.7b. Dashed gray line trace along the CNP at various V_{SL}. (b) R_{yy} at CNP as a function of V_{SL} at various temperatures.

4.4 R_{xx} and R_{yy} at $B \neq 0T$

Magnetotransport response of a graphene 2DSL system consists of the main Dirac fan as expected in pristine graphene, which consists of quantum Hall and Shubnikov-de Haas (sdH) oscillations, and several side fans related to the satellite Dirac cones. In addition to these features, graphene 1DSL also shows commensurability oscillations in its magnetoresistance R_{xx} and R_{yy}.

Figure 4.11(a)(b) shows R_{xx} and R_{yy} in the $L = 47nm$ device, as a function of vertical magnetic field B and carrier density n, respectively. In these measurements $V_{SL} = 48V$, corresponding to $u \approx 3\pi$. R_{xx} versus magnetic field shows a Landau fan of integer quantum Hall states emanating from the CNP at $n = 0$ with filling factors identical to those of pristine graphene ($v = 4N + 2$, N integer). In addition, two satellite fan-like features are also visible, emanating from $n = \pm 2.7 \times 10^{11} \, cm^{-2}$ (Fig. 4.11a, red arrows), the carrier densities at which satellite Dirac cones emerge as predicted by band structure simulation.

R_{xx} minima features emanating from the $n = -2.7 \times 10^{11}\ cm^{-2}$ satellite fan are indicated in the fan tracing diagram shown in Fig. 4c. The existence of side Landau fans is a further confirmation that band structure modification has happened in out graphene 1DSL system. Notice that the side fans in graphene 1DSL, unlike the case in graphene 2DSL, are not related to "Hofstadter butterfly"[1, 2, 6], as it is not related to the cancellation of Aharonov-Bohm phase across several SL unit cells.

In contrast, side fans are missing in R_{yy}. We can interpret this as an overlap of the transport channel from the Landau levels of the satellite Dirac cones and that from the open orbits. Due to the magnetic breakdown across the Brillouin zone boundaries of 1DSL, the first bands beyond the van Hove singularity can no longer form well-quantized orbits. Instead, they make open orbits[7] which have nearly continuous energy levels and small conductivities under magnetic fields.

Along the direction parallel to the SL basis vector, the magnetoconductivity ($R_{xx} \propto \sigma_{yy}$) from the Landau levels of the satellite Dirac cones is much higher than that from the open orbits, due to the finite energy dispersion of the Landau levels. Thus, we can clearly see the side fans in R_{xx}. In R_{yy} ($\propto \sigma_{xx}$), on the other hand, the contribution from each channel, either from each open orbit or from each Landau level of the satellite Dirac cones, is infinitesimal since the wavefunctions under B are localized along x. Note that, although σ_{xx} from the satellite Dirac cones is quite small, it is not completely zero since there are small inter-Landau level contributions, i.e., $\langle n|v_x|n'\rangle \neq 0$ for $n \neq n'$. Thus, the conductivity σ_{xx} mainly scales with the density of levels. Since the energy spacing between the open orbits is much smaller than that between the Landau levels of the satellite Dirac cones, the conductivity from the side fans is buried to that from the open orbits. This is the reason why we cannot see side fans in R_{yy}.

Note that, unlike the fans from the satellite Dirac cones, we can clearly see the Landau fans from the main Dirac cone in R_{yy}, since the first bands below the van Hove singularity form well-quantized orbits.

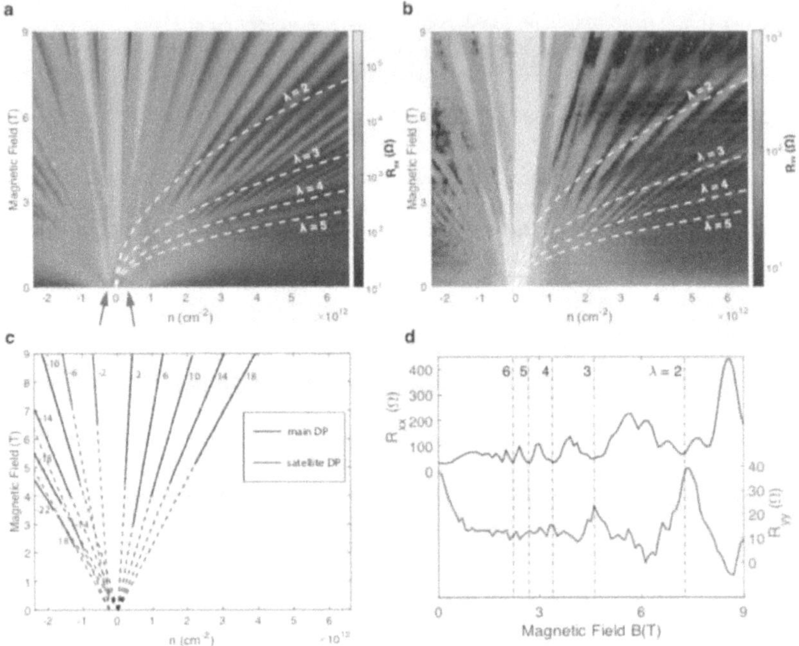

Figure 4.11 Magnetotransport in the $L = 47nm$ **graphene** 1DSL device (a) Measured longitudinal resistance R_{xx} as a function of carrier density n and magnetic field B, in a $L = 47nm$ device with $V_{SL} = 48V$. Red arrows indicate the density locations of the where satellite Landau fans converge. (b) R_{yy} measured from the same devices. In **a** and **b** dashed white lines trace along $2r_c = \left(\lambda - \frac{1}{4}\right)L$, with λ being an integer and $r_c = \hbar\sqrt{\pi n}/eB$ the cyclotron radius (see text). (c) Fan tracing diagram highlighting traces of R_{xx} minima from Fig. 4a. The R_{xx} minima features associated with the main (satellite) Dirac point are colored in **black** (red), respectively. Numbers indicate the filling fractions associated with the main or

satellite fan. (d) R_{xx} and R_{yy} as a function of magnetic field B at carrier density $n = 6.53 \times 10^{12} cm^{-2}$.

Dashed lines indicate the magnetic field at which the corresponding oscillation is theoretically expected (see text). In order to suppress the quantum Hall oscillations and highlight the commensurability oscillations, the plotted curves are obtained by averaging the measured resistance over a small density window of $(6.53 \pm 0.23) \times 10^{12} cm^{-2}$.
It is also worth noticing that the magnitude of the magnetoresistance R_{xx} is about 2 orders higher than R_{yy}, a sign of transport anisotropy. In a crude approximation, the conductivities σ_{xx} and σ_{yy} are proportional to the square of the expectation value of v_x and v_y, respectively. At moderate magnetic field strength B, the SL potential $V(x)$ lifts the degeneracy of LLs and give an energy dispersion along k_y. This makes the intra-Landau level contribution to v_y ($\langle n|v_y|n\rangle$, where n is the Landau level index) finite, while that to v_x ($\langle n|v_x|n\rangle$) vanishes since the wavefunction under B is localized along x [8, 9]. Thus, we expect $\sigma_{yy} \gg \sigma_{xx}$. The magnetoresistance $R_{xx} = \sigma_{yy}/(\sigma_{xx}\sigma_{yy} - \sigma_{xy}\sigma_{yx})$ can be approximately expressed as $R_{xx} \sim \sigma_{yy}/\sigma_{xy}^2$, since $|\sigma_{xx}\sigma_{yy}| \ll |\sigma_{xy}|^2$ in our system. Likewise, $R_{yy} \sim \sigma_{xx}/\sigma_{xy}^2$. Thus, the theoretically predicted $\sigma_{yy} \gg \sigma_{xx}$ is consistent with the experimentally observed $R_{xx} \gg R_{yy}$. Such dramatic asymmetry between R_{xx} and R_{yy} cannot be observed either in the conventional 1DSL with a weak potential V and a long superlattice period L or in the 2DSL. In 1DSL, $\langle n|v_y|n\rangle$ is roughly proportional to the energy width ΔE of LLs and inversely proportional to the SL Brillouin zone size along k_y, Δk_y. Since ΔE is proportional to V, and $\Delta k_y = L/l_B^2 = eBL/\hbar$, where $l_B = \sqrt{\hbar/eB}$ is the magnetic length, the anisotropy between the magnitude of R_{xx} and R_{yy} in 1DSL approximately scales with V/L. And in usual 2D SL with $L_x \sim L_y$, the magnetic minibands exhibit energy dispersion in both k_x and k_y directions, aka Hofstadter butterfly. Thus, $\langle n|v_x|n\rangle$ no longer vanishes, and R_{yy}

becomes comparable to R_{xx}. Thus, the 1DSL developed in this dissertation, which has a strong potential V and a very short period L, provides a unique opportunity to show the dramatic anisotropy between the magnetotransport R_{xx} and R_{yy}.

At high enough carrier density n, both R_{xx} and R_{yy} show curved features roughly following $B \sim \sqrt{n}$. White dashed lines in Fig. 4.11a and b trace along (B,n) pairs satisfying the condition $2r_c = \left(\lambda - \frac{1}{4}\right) L$, and these lines follow neatly with the curved features at high n. It is also evident that for R_{xx}, the white dashed lines trace along R_{xx} minima while for R_{yy}, maxima. Figure 4.11d shows R_{xx} and R_{yy} as functions of magnetic field B at fixed carrier density $n = 6.53 \times 10^{12} cm^{-2}$. To suppress contributions from quantum Hall oscillations that may mask the commensurability oscillations[8, 10, 11], the plotted curves are actually obtained by averaging the measured resistance over a small density window of $(6.53 \pm 0.23) \times 10^{12} cm^{-2}$. In our measurements, the data resolution in carrier density is $\Delta n = 3.34 \times 10^{10} cm^{-2}$. Therefore $\frac{1}{15}\sum_{k=-7}^{7} R_{xx}(n + k\Delta n, B)$ and $\frac{1}{15}\sum_{k=-7}^{7} R_{yy}(n + k\Delta n, B)$ are the quantities being plotted in Figure 4.11d. Figure 4.12a shows the R_{xx} and R_{yy} at $n = 6.53 \times 10^{12} cm^{-2}$ without taking any average over a neighboring n range. While we can still see R_{xx} minima and R_{yy} maxima happening at the predicted magnetic fields values satisfying $2r_c = \left(\lambda - \frac{1}{4}\right) L$ marked by dashed lines, there are obscured by the more prominent and frequent quantum Hall oscillations.

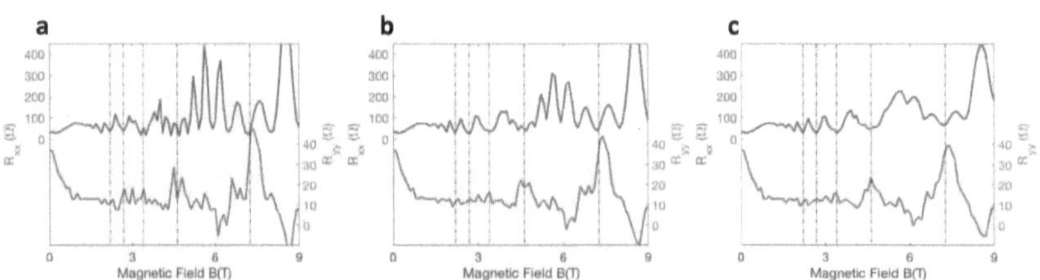

Figure 4.12 Averaging over a small window of carrier density n to smooth out quantum Hall oscillations (a) R_{xx} and R_{yy} as a function of magnetic field B at $n =$

$6.53 \times 10^{12} cm^{-2}$ without taking average values in a neighboring carrier density range, for the L=47nm graphene 1DSL device. (b) The average R_{xx} and R_{yy} from the carrier density range $n = (6.53 \pm 0.13) \times 10^{12} cm^{-2}$. (c) The average R_{xx} and R_{yy} from the carrier density range $n = (6.53 \pm 0.23) \times 10^{12} cm^{-2}$, same as Figure 4.11d. Dashed lines indicate magnetic fields at which commensurability oscillation minima in R_{xx} (and thus maxima in R_{yy}) are predicted to occur based on Equation 2 in the main text.

Commensurability oscillations can be understood both semi-classically and quantum mechanically. In the semiclassical picture, the guiding centers of the cyclotron orbits of carriers in 1DSL drift along the y direction. Drifting velocity is proportional to the local electric field E(x), which is modulated by the 1DSL potential. It follows that drifting velocity is enhanced or reduced depending on whether $E(X + r_c)$ and $E(X - r_c)$ have the same or opposite sign, where X is the x-coordinate of the guiding center. Further analysis shows that R_{yy} minima and R_{xx} maxima are achieved when the resonance condition $2r_c = \left(\lambda + \frac{1}{4}\right)L$ (notice the plus sign instead of minus) is satisfied[12].

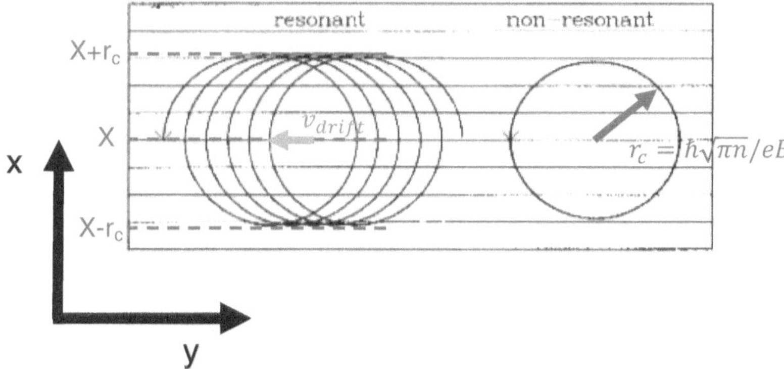

Figure 4.13 Semiclassical explanation of commensurability oscillation Reprinted from [12]. According to Beenakker, in a 1DSL system under vertical magnetic field, the guiding

center of the cyclotron orbits tends to move along the y axis. The drift velocity is greatest

when $2r_c = \left(\lambda + \frac{1}{4}\right) L$ is satisfied.

Alternatively, commensurability oscillations can be understood quantum mechanically by a $1/B$-periodic oscillation in the Landau level bandwidth, which reaches zero when $2r_c = \left(\lambda - \frac{1}{4}\right) L$ is satisfied. We note that the commensurability oscillations along the two measured directions appear out of phase such that when R_{xx} fades, R_{yy} grows, and vice versa. The Landau level degeneracies are lifted by the 1DSL potential with each Landau level acquiring a dispersion in k_y only. Thus, when the LL width in the direction of k_y is large, R_{xx} is large, since $R_{xx} \propto v_y^2$ where $v_y = \frac{1}{\hbar}\frac{dE}{dk_y}$. Conversely, $R_{yy} \propto v_x^2$ and so is less sensitive to variation of the intraband scattering and instead varies directly with the DOS, causing R_{yy} to scale inversely with the level width. Therefore, zero LL bandwidth (zero dispersion in k_y, corresponding to the condition in (2)) leads to an R_{xx} minimum, simultaneous with an R_{yy} maximum.

Figure 4.14 shows commensurability oscillation in R_{xx} as three different temperatures. We notice that the commensurability oscillation envelope persists at T=20K while quantum Hall oscillation have been largely smoothed out. This agrees with Drienovsky's measurement.[11]

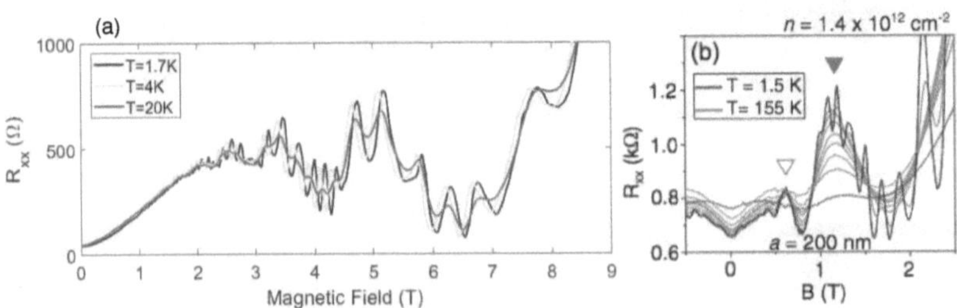

Figure 4.14 Temperature dependence of commensurability oscillation in R_{xx} (a) R_{xx} as a function of magnetic field B at fixed carrier density $n = 7.97 \times 10^{12} cm^{-2}$ for three

different temperatures. $L = 55nm$, $V_{SL} = 76V$. (b) R_{xx} versus B at $n = 1.4 \times 10^{12} cm^{-2}$ at various temperatures in Drienovsky's measurement. Reprinted from [11].

4.5 Summary

Two graphene 1DSL devices with pitch L=47nm and L=55nm respectively are made and achieved higher carrier mobilities than previous devices. These devices are L-shaped, enabling measurements of transport anisotropy.

The R_{xx} and R_{yy} measurements at zero magnetic field reproduces the calculated values in Figure 2.8 and are strong evidence for the periodic flattening/unflattening of main and 1st satellite Dirac cones. R_{xx} and R_{yy} are very symmetric about n=0 but not symmetric enough about $V_{SL} = 0$, suggesting that the true 1DSL potential has even symmetry but not odd symmetry. At high temperature an "X"-shaped R_{xx} feature still exists, and can be explained by Drienovsky's modelling of 1DSL as a series of pn junctions. Magnetotransport data in R_{xx} show satellite Landau fans that further corroborate the existence of band structure modification. Commensurability oscillation in R_{xx} and R_{yy} are also seen.

Chapter 5. Plasmonic band structure engineering in a graphene 2D superlattice system

5.1 Surface plasmon polaritons (SPPs)

In this chapter I discuss my research efforts related to the band structure engineering of surface plasmon polaritons (SPPs).

Just like phonons that are the quasiparticles of mechanical vibrations of atoms in a crystal, plasmons are quasiparticles of plasma oscillation of electrons. Surface plasmons can be excited by light at a metal-dielectric interface. This hybrid excitation of light and mobile electrons is called surface plasmon polariton.

The origin of plasma oscillation in solids can be described by the following model: Consider a chain of N equally spaced atoms with each electron displaced from the corresponding nucleus by a distance of δx. This displacement introduces an electric field $E_s = Ne(\delta x)/\epsilon_0$ pointing in the direction of the electron displacement, which pulls the electrons back to their original positions. Because of their inertia, the electrons will overshoot and oscillate around their equilibrium positions with a characteristic frequency known as the plasma frequency. Suppose at any given time t, the electron displacement is given by $\delta x = \delta x_0 \exp\left(-i\omega_p t\right)$, plug into the equation of motion $m\frac{d^2(\delta x)}{dt^2} = (-e)E_s$ we have the plasma frequency $\omega_p = \sqrt{Ne^2/m\epsilon_0}$. In Drude model, dielectric constant in metal is given by $\epsilon(\omega) = \left(1 - \frac{\omega_p^2}{\omega^2}\right) + i\frac{\omega_p^2\gamma}{\omega^3}$ where γ is a damping factor. Note that the above discussion assumes electrons to be "free", and the electron is not bound to a nucleus by virtue of its atomic structure. In certain materials such as gold, we need to take contributions from bound electrons into account when calculating the dielectric constant $\epsilon(\omega)$.

In bulk metal the dispersion relation of electromagnetic waves can be found by $k^2 = |\mathbf{k}|^2 = \epsilon\omega^2/c^2$. Ignoring the imaginary part of ϵ that approaches zero at large ω, we get

$$\omega(k) = \sqrt{\omega_p^2 + \frac{k^2}{c^2}} \tag{1}$$

However, this is a longitudinal excitation mode with $\boldsymbol{k} \parallel \boldsymbol{E}$ that cannot be excited by light, a transverse wave.

Another type of plasmon polariton is the surface plasmon polaritons that propagate along the interface between a metal and a dielectric medium. Consider the following ansatz for the electromagnetic fields in metal and dielectric near such an interface:

$$\begin{cases} \boldsymbol{H}_d = \left(0, H_{yd}, 0\right)e^{i(k_{xd}x+k_{zd}z-\omega t)} \\ \boldsymbol{E}_d = (E_{xd}, 0, E_{zd})e^{i(k_{xd}x+k_{zd}z-\omega t)} \end{cases} \text{for } z > 0 \text{ (in dielectric)} \tag{2}$$

$$\begin{cases} \boldsymbol{H}_d = \left(0, H_{ym}, 0\right)e^{i(k_{xm}x+k_{zm}z-\omega t)} \\ \boldsymbol{E}_d = (E_{xm}, 0, E_{zm})e^{i(k_{xm}x+k_{zm}z-\omega t)} \end{cases} \text{for } z < 0 \text{ (in metal)} \tag{3}$$

At the interface, $E_{xm} = E_{xd}$, $H_{ym} = H_{yd}$, and $\epsilon_m E_{zm} = \epsilon_d E_{zd}$. Plug in (2)(3) to the interface boundary conditions and Maxwell's equations, we get $k_{xm} = k_{xd} = \frac{\omega}{c}\sqrt{\frac{\epsilon_m \epsilon_d}{\epsilon_m + \epsilon_d}}$. Unifying the notations k_{xm} and k_{xd} into k_x, and noting that $\epsilon_m(\omega) = 1 - \omega_p^2/\omega^2$ in metal if we ignore the damping factor γ, the dispersion relation for surface plasmons can be described as $k_x = \frac{\omega}{c}\sqrt{\frac{(\omega^2-\omega_p^2)\epsilon_d}{(1+\epsilon_d)\omega^2-\omega_p^2}}$. Notice that if we consider the imaginary part of $\epsilon_m(\omega)$ due to the damping factor γ then k_x, just like k_{zd} and k_{zm}, carries an imaginary part and therefore SPP propagation attenuates exponentially along the x direction.

Figure 5.1 shows the dispersion relations of photon, surface plasmon, bulk (volume) plasmon and their polaritons. Although unlike bulk plasmon polaritons, SPPs are a transverse wave, they do not have the same dispersion relation as photons and at a given frequency ω there is always a nonzero mismatch in wave vector between photon and SPP that this photon is supposed to excite. Total internal reflection, diffraction gratings and near-field optics can be used to bridge the wave vector mismatch.

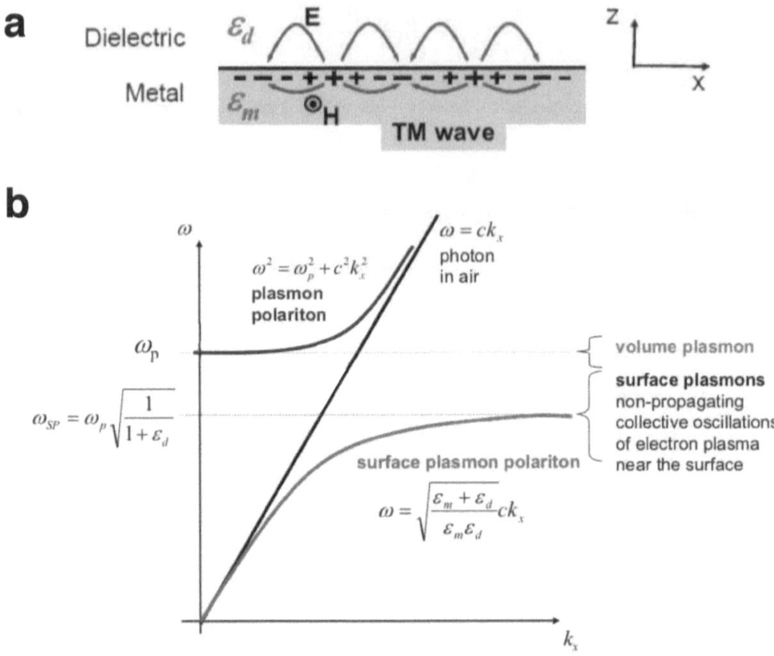

Figure 5.1 Surface plasmon polaritons (SPPs) and its dispersion relations Reprinted from [1] (a) Schematics of electric field and charge distribution due to SPP at a dielectric-metal interface (b) Dispersion relations of bulk plasmon polaritons (purple), surface plasmon polaritons (red), bulk plasmons that are not coupled to light (green dashed), surface plasmons that are not coupled to light (black dashed), and photons (**black solid**), assuming damping factor $\gamma = 0$.

Scattering-type scanning near-field optical microscopy (s-SNOM) is a technique that efficiently excites and detects SPPs. In SNOM, the excitation laser is focused through an aperture with a diameter smaller than the laser wavelength, resulting in an evanescent field that contains Fourier components whose wave vectors are larger than that of the incoming laser. Laser beam #1 (yellow arrows in Figure 5.2) is directly backscattered by the metalized

AFM tip with radius ~30nm and positioned ~30nm above the sample device and registered

on the detector, while laser beam #2 (red arrows in Figure 5.2) first reaches a metal launcher

with thickness ~60nm, excites plasmons in the metal-dielectric interface which travels to and

scatters on the AFM tip as well. Both the scattering on the AFM tip and the propagation of

light across the metal launcher edge causes diffraction that bridges the wave vector mismatch.

The two beams interfere on the detector with a phase difference of $\Delta\phi = 2\pi x/\lambda_p$ where x is

the extra distance travelled by beam #2 on the sample surface. This phase difference causes

oscillatory fringes in the near-field image $s(\boldsymbol{r}, \omega)$, which measures $|E_z|$ directly above the

sample surface, with wavelength λ_p.

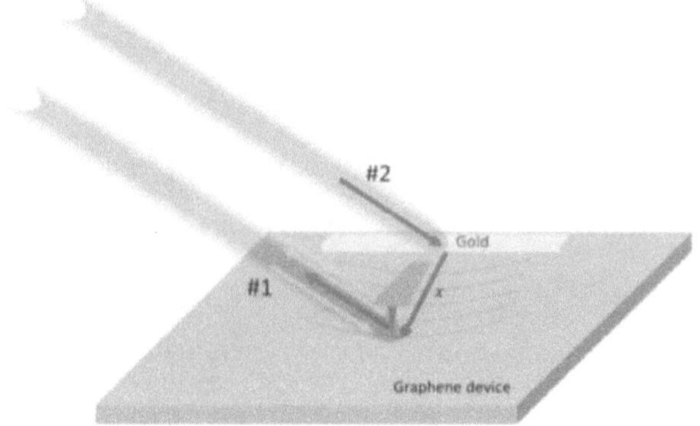

Figure 5.2 Detection of SPP fringes Reprinted from [2]. The detected signal is the

interference between the scattered Beam #1 (yellow) and Beam #2 (red). The phase shift

between the two beams induces oscillatory fringes in the near-field image $s(\boldsymbol{r}, \omega)$ with

wavelength λ_p.

Because the wavelength of SPP is shorter than that of light at the same frequency, controlling

the propagation of SPP modes effectively controls the propagation of light at a length scale shorter than its wavelength. Therefore, SPPs have recently received considerable interest in the field of opto-electronics.

5.2 Plasmonic band structure of a graphene 2DSL system

2D materials have received considerable interest for the studying of SPPs due to long SPP propagation lengths[3], the ease of carrier density tuning by gating[4], and the large SPP confinement ratios λ_0/λ_p in these materials. For graphene the confinement ratio can be as high as 66. [3]

As a result of the isotropic optical conductivity tensor in graphene, graphene SPPs have a circular wavefront.(Figure 5.3d) [5]. By solving Maxwell's equations and imposing boundary conditions just like what we did for the metal-dielectric interface model in Section 5.1, we can calculate the SPP dispersion relation in a graphene sheet sandwiched by two semi-infinite thick hBN slabs.[6] (Figure 5.3a) The dispersion relation has two band gaps at 90~100meV and 150~190meV respectively, where no   SPP modes are allowed. In a realistic device the thicknesses of the encapsulating hBN flakes are finite. In particular, the top hBN needs to be thinner than ~10nm such that the AFM tip in the s-SNOM system can be as close to the top hBN-graphene interface as possible in order to boost signal-to-noise ratio. Figure 5.3b shows the calculated SPP dispersion relationship in encapsulated graphene with top BN thickness 7nm and bottom BN thickness 46nm. As a result of thin film effects, certain phonon polariton modes exist in the energy regimes labelled by orange.

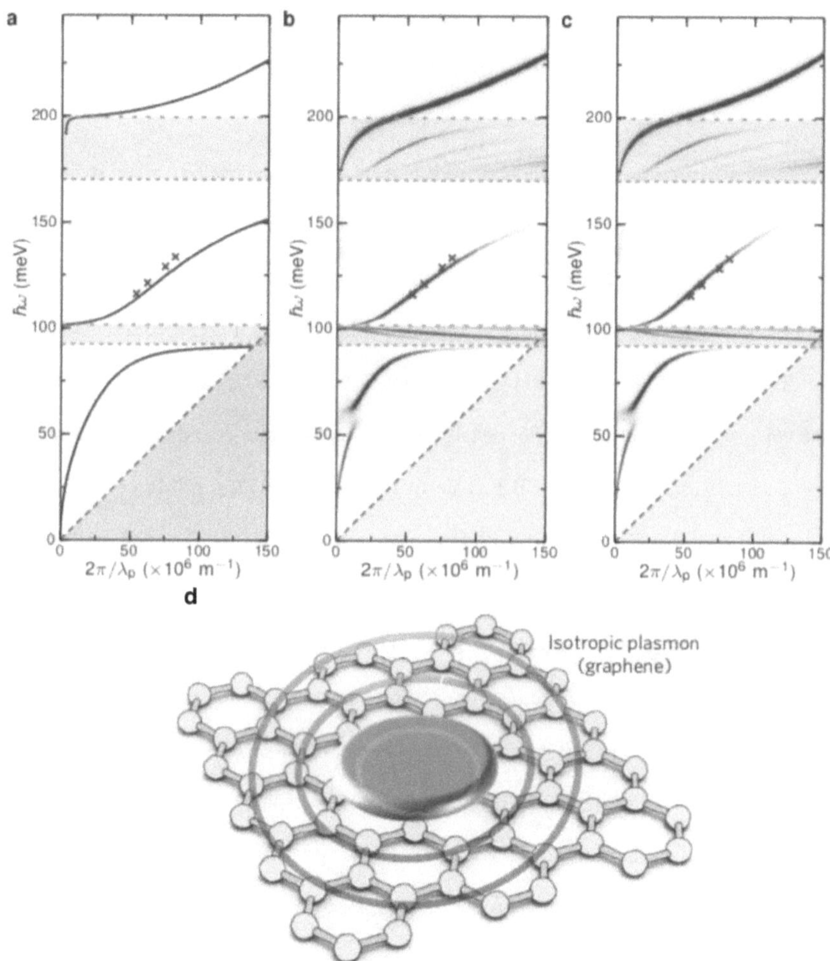

Figure 5.3 SPP dispersion relations in graphene (a) SPP dispersion in hBN encapsulated graphene as calculated by the Drude model, assuming that the hBN flakes are semi-infinite in the direction perpendicular to graphene (b) Same as (a) but taking thin film effects into account with top hBN thickness = 7nm and bottom hBN thickness = 46nm (c) Same as (b) but taking nonlocal conductance into account. For (b) and (c) certain modes in the orange regime are phonon polariton modes and unrelated to SPP. (a~c) reprinted from [6]. (d)

Schematics of SPP wavefront launched from a metal launcher disk on top of graphene, showing the isotropic nature of SPP in graphene. Reprinted from [5].

The rotational symmetry of the graphene SPP dispersion relation can be broken by imposing a 2D dielectric superlattice consisting of an array of pillars. Furthermore, the Si gate voltage V_g tunes both the average carrier density in graphene $\bar{n}$ and the amount of carrier density modulation between the pillar regions and the non-pillar regions. Figure 5.4K shows the three-dimensional SPP dispersion in graphene under 2DSL as a function of $\boldsymbol{k}$ and $\bar{n}$, as long as the vertical cut along fixed carrier density $\bar{n} = 4.1 \times 10^{12} cm^{-2}$ and the horizontal cut along $\omega = 890 \; cm^{-1}$. Figures 5.4B,E,H shows more cuts along three different values of $\bar{n}$. At $\bar{n} = 3.6 \times 10^{12} cm^{-2}$ (Fig 5.4B) SPP modes are allowed near the M direction of the superlattice Brillouin zone at $\omega = 890 \; cm^{-1}$, and the simulated $|E_z|$ map (Fig. 5.4D) predicts SPP propagation along the M direction but not in the K direction. At $\bar{n} = 4.8 \times 10^{12} cm^{-2}$ (Fig 5.4E) SPP modes are allowed near the K direction at $\omega = 890 \; cm^{-1}$, and the simulated $|E_z|$ map (Fig. 5.4G) predicts SPP mode propagation along the K direction but not the M direction. At $\bar{n} = 6.0 \times 10^{12} cm^{-2}$ (Fig. 5.4H), SPP modes are allowed along both the M and K directions, and the simulated $|E_z|$ map (Fig. 5.4J) predicts a mostly isotropic SPP propagation map as seen in pristine graphene. (Fig. 5.3d). Therefore, the presence of 2DSL has engineered an anisotropic band gap in the dispersion relation of SPP that allows us to control the propagation direction of SPP solely by adjusting V_g, which controls $\bar{n}$. From now on we use the terms "dispersion relation" and "band structure" interchangeably in the context of SPPs but bear in mind that unlike elections, polaritons in general are bosonic quasiparticles and thus Pauli exclusion is not applicable. Therefore, there is no "Fermi sea" in SPPs and the SPP band structure merely tells us which modes are allowed and which are not.

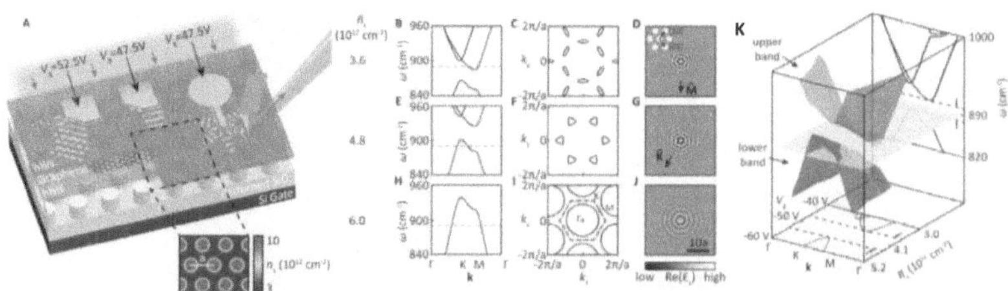

Figure 5.4 SPP dispersion relation in graphene 2DSL system (A) Schematics of a

graphene 2DSL device. Inset shows the carrier density modulation between the pillar and

non-pillar regions (B,E,H) SPP dispersion sliced along three different values of average

carrier densities $\bar{n}$. (C,F,I) Equi-energy contours in the k-space along $\omega = 890\ cm^{-1}$ at

three different values of $\bar{n}$. (D,G,J) simulated $|E_z|$ map in real space for SPP launched from

the center, at three different values of $\bar{n}$. (K) SPP dispersion as a function of k and $\bar{n}$, and

the slice along a fixed frequency (red) and a fixed $\bar{n}$ (**black**).

5.3 A previous study on graphene 2DSL plasmonics

In 2018 Xiong et al investigated the propagation of SPPs along K direction in a graphene

2DSL device.[2] Two mirroring regions of triangular superlattice are separated by a one-

dimensional domain wall composing of two parallel arrays of pillars. The 2DSL has 85nm

pitch and its plasmonic dispersion relation is shown in Figure 5.5e. The laser frequency and

thus the SPP frequency is fixed at $\omega = 904\ cm^{-1}$. At $V_g = -90V$, according to the

calculated dispersion relations (Fig. 5.5e, red curves), SPP propagation is allowed in both M

and K directions, and near field images confirm the existence of SPP fringes along the K

direction (Fig 5.5c,d). At $V_g = -40V, -60V$ and $-70V$, SPP fringes are very faint and, at

least in the case of $V_g = -60V$, SPP propagation in the 2DSL region along the K direction

(Fig. 5.5d, purple curve) is suppressed compared to in the pristine graphene region (Fig. 5.5d,

purple dashed curve). Near the SPP band gap around $V_g = -65V$, no SPP modes are allowed

to propagate along any direction in the regions subject to 2D triangular superlattice potential, and only propagation along the 1D domain wall is allowed, as seen in the very bright near field signal along the domain wall at $V_g = -70V$. (Fig. 5.5c)

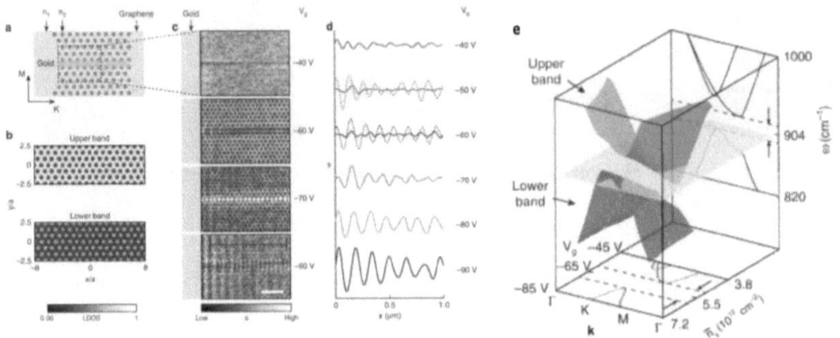

Figure 5.5 Gate-tunable plasmonic response in a graphene 2DSL device along the K direction Reprinted from [2]. (a) Schematics of a photonic crystal structure with pitch 80nm and an artificial domain wall in the middle. K and M directions are labelled. (b) Simulated local density of state (LDOS) maps for the upper and lower plasmonic bands. (c) Experimental near field images $s(\mathbf{r}, \omega)$ acquired at different V_g. Scale bar: 400nm. (d) Line profiles of measured $s(\mathbf{r}, \omega)$ taken in a photonic crystal region away from the domain wall. Dotted lines indicate the corresponding line profiles in the unpatterned region (e) Plasmonic dispersion relations of a graphene 2DSL system with pitch 80nm along with its slices along $\omega = 904\ cm^{-1}$ (red) and $V_g = -65V$ (**black**).

5.4 Device fabrication

Despite these early signs of gate-controlled SPP propagation in this graphene 2DSL device, it is unable to measure SPP propagation along the M direction due to the extremely limited amount of usable device area available in the M direction of the metal launcher. The most challenging part of fabricating a graphene 2DSL device is the making of the hBN-graphene-hBN stack, a stack composing of three layers of thin (<10nm) 2D materials while maintaining both high device quality (no breaks, folds, wrinkles, interlayer bubbles and few polymer residues) and a relatively large device area. The top hBN needs to be thin to maximize the detected near field signal, and the bottom hBN also needs to be thin to avoid screening out the carrier density modulation in graphene due to the superlattice pillars. Thin layers of 2D materials tend to break, fold, wrinkle much more easily than their thicker counterparts during the van der Waals transfer process, causing defects in the assembled heterostructure. Polypropylene carbonate (PPC) does not tend to easily stick to thin layers of 2D materials and it is very challenging to pick up the topmost layer of our stack using a PPC slide, as what we did in the graphene 1DSL device in Chapter 4. Since the bottom hBN is also thin, it does not help to reverse the order of assembly, start from the bottom layer and flip the stack. Using other polymers such as polycarbonate (PC) that picks up thin flakes more easily than PPC does for stack assembly is also not an option as they introduce too much residue on the device surface even after vacuum annealing.

Xiong et al's device in 2018[2] fabricated the stack by first picking up a thick (>20nm) layer-of hBN using PPC. Then, the top thin hBN is positioned between the PPC touchdown point and the picked-up thick hBN flake, with partial overlap between the thin and thick hBN flakes. As the PPC wavefront is slowly pulled back, the thick hBN, already on the PPC, would lift the thin hBN flake up thanks to the overlapping region. Once the entire thin hBN flake is picked up, the graphene and the bottom hBN flake can also be picked up due to van der Waals forces. I improved this method by pre-patterning the thick hBN by etching a hole with 7μm~12μm diameter. After this pre-patterned thick hBN flake is picked up using PPC, thin flakes are positioned such that the thick hBN hole lies inside the thin flake entirely. Because the thin flakes now feel van der Waals attraction to the thick hBN everywhere outside the hole, the transfer process is generally smoother and tends to produce larger usable areas compared to the 2018 method.

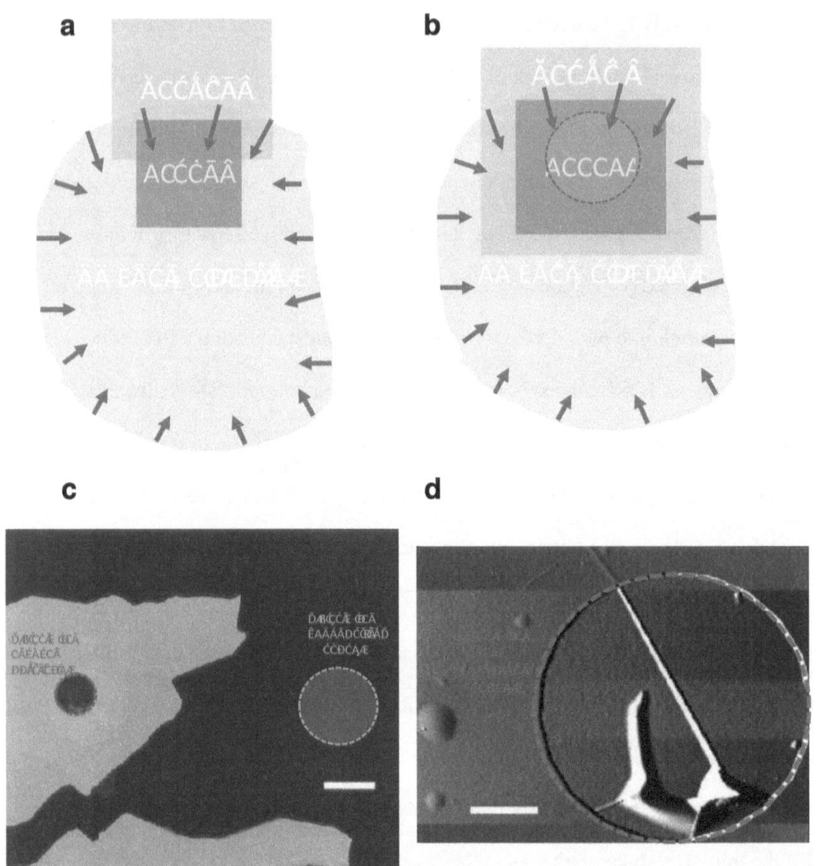

Figure 5.6 Stacking a thin hBN-graphene-thin hBN stack (a,b) Schematics of the 2018 and new methods of thin stack assembly, respectively. Blue arrows indicate that the PPC/SiO₂ interface region contracts during the pick up process. © Optical microscope image of a thin hBN-graphene-thin hBN stack made using the new method after being transferred onto a SiO₂/Si substrate with 2DSL pillars. Scale bar=10μm. Disks labelled by orange indicate regions with 2DSL pillar inside. The disk labelled by blue in b,c,d indicate the etched hole in the top thick hBN inside which the thin hBN-graphene-thin hBN stack exists. (d) AFM amplitude error image of the stack shown in (c). Besides a fold that bisects the thin

stack region and a bubble on the lower-left, the stack is very clean and smooth. Scale bar=2μm

It is also worth noting that for both the 2018 device and my new device, h^{11}BN which is isotopically pure in boron, is used instead of the usual hBN. Boron has two naturally occurring stable isotopes, ^{10}B (abundance 20%) and ^{11}B (abundance 80%) with different atomic mass. It has been shown experimentally [7] that for phonon polaritons propagating on an hBN surface, the propagation length increases from ~4μm in hBN with natural boron abundance to >10μm in hBN with 99.2% ^{11}B abundance. Although no studies have been made on the effect of isotopically purity of boron in hBN on SPP propagation, we have decided to use h^{11}BN flakes for both the top and bottom thin hBN in an attempt to maximize SPP propagation length.

The fabrication of 2DSL pillars on a SiO_2/Si substrate is very similar to the fabrication of 1DSL lines explained in Section 4.1.2. To make pillars, regions inside hexagonal tiles shown in Figure 5.7a has their top ~50nm of SiO_2 etched away, leaving the regions outside these tiles as SiO_2 pillars. Because Nanobeam e-beam lithography system only recognizes displacement up to 1nm, special care is taken such that the origin of each SL unit cell has its x and y coordinates rounded to the nearest integer nanometer to ensure that the resulting pillars have homogenous pillar diameters.

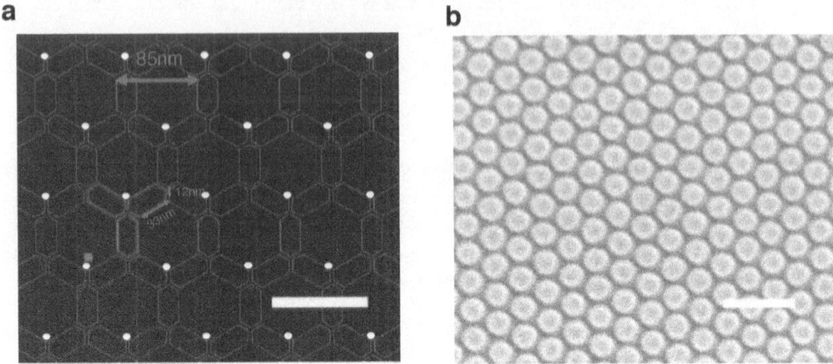

Figure 5.7 Fabrication of 2DSL pillars (a) AutoCAD design of the array of hexagon regions marked in red that would have their top ~50nm of SiO_2 etched away, leaving pillars behind. The unit cell of this array is highlighted. The origin of each unit cell, marked by the white dot, have their x and y coordinates rounded to the nearest integer nanometer. Scale bar=100nm (b) SEM image of a finished 2DSL pillar array. Gray region: Pillars. Scale bar=200nm

After both the stack and the 2DSL are fabricated, the stack is transferred onto 2DSL and at least 2 metal-to-graphene electric contacts are made. Two contacts are necessary to determine the gate voltage offset in the device by finding the V_g at which the two-point resistance of graphene reaches maximum. Finally, metal launchers shaped as seen in Figure 5.4A are deposited on top of the stack to allow for SPP propagation in directions other than K.

5.5 Measurement results

Figure 5.8C,D,E shows near field signal from a graphene 2DSL pillar device at $\omega = 890\ cm^{-1}$ with $V_g = 47.5V, 52.5V, 72.5V$, respectively. The SPPs are launched from a pentagon shaped launcher that have two edges 30° apart, allowing simultaneous launching in both K and M directions. Figure 5.8F shows the line profiles along the K and M directions acquired from Figure 5.8 C,D,E. At $V_g = 47.5V$, SPP fringes appear in the M direction but not in the K direction, agreeing with the calculation that SPP has modes near M but not near K at $\bar{n} = 3.5 \times 10^{12} cm^{-2}$ (Figure 5.8B). At $V_g = 52.5V$, SPP fringes appear in the K direction but not in the M direction, agreeing with the calculation that SPP has modes near K but not near M at $\bar{n} = 4.7 \times 10^{12} cm^{-2}$ (Figure 5.8B). At $V_g = 72.5V$, SPP fringes appear in both K and M directions, agreeing with the calculation that SPP has both modes near K and modes near M at $\bar{n} = 6.0 \times 10^{12} cm^{-2}$. Therefore, V_g acts as a tunable switch that can turn SPP transport in K and M directions on and off.

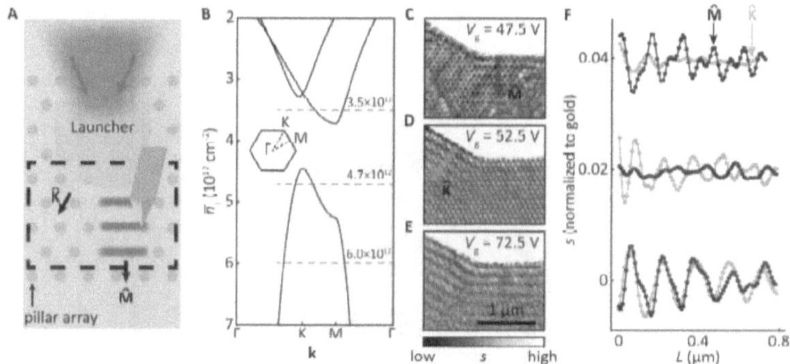

Figure 5.8 Gate bias as a SPP switch (A) Schematics of the device with the scanned region in the dashed box.(B). Plasmonic band structure of graphene subject to a 2DSL pillar superlattice with pitch L=85nm, sliced along constant frequency $\omega = 890\ cm^{-1}$. (C,D,E) Near field signal at $V_g = 47.5V$ ($\bar{n} = 3.5 \times 10^{12} cm^{-2}$), $V_g = 52.5V$ ($\bar{n} = 4.7 \times 10^{12} cm^{-2}$) and $V_g = 72.5V$ ($\bar{n} = 6.0 \times 10^{12} cm^{-2}$), respectively. (F) Averaged line profiles along K and M directions from the near field images in (C,D,E).

By modifying the shape of the launcher into a needle, SPPs are able to launch into almost any direction in the xy plane. The cropped near-field images near the launcher apex are Fourier transformed using a Hann window and then symmetrized to get an image in reciprocal space.

(Figure 5.9C,F,I) More specifically, we consecutively rotated the raw Fourier images in 60°

intervals, leading to 6 transformed images. We then applied the mirror operation to these 6

transformed images, resulting in 6 additional images. All 12 transformed images were

averaged and result in the Fourier images shown in Figure 5.9C,F,I. Figure 5.9C shows

Fourier transform maxima around M point. Figure 5.9F shows Fourier transform maxima

around K point. Figure 5.9I shows a mostly isotropic Fourier transform map. These Fourier

transform maps can be seen as equi-energy contours in the graphene 2DSL SPP band

structure, as they fit neatly into the calculated SPP band structure in Figure 5.9K. If we

measure near field images at more values of V_g we can effectively probe and map the entire

SPP band structure of the device.

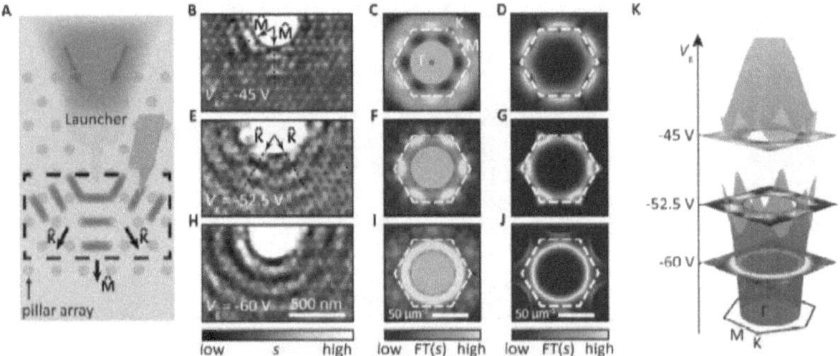

Figure 5.9 Fourier analysis of polaritonic images A. Device schematics showing needle-

like launcher. B,E,H. Near field images taken for an 80nm pitch graphene 2DSL pillar device

at three different gate biases. At $V_g = -45V$ (B) SPP propagation is predominately along

the M direction. At $V_g = -52.5V$ (E) SPP propagation is predominately along the K

direction. At $V_g = -60V$ (H) SPP propagation is mostly isotropic. (C,F,I) Symmetrized

Fourier transform of (B,E,H) respectively. (D,G,J) Simulated Fourier transform of near field

images at $V_g = -45V, -52.5V, -60V$, respectively. (K) Calculated SPP band structure for

an 80nm pitch graphene 2DSL pillar device with Fourier transformed images (C,F,I)

overlayed at the corresponding V_g values, showing that the Fourier transform of near-field images can be understood at equi-V_g contours of SPP band structure.

5.6 Summary

Surface plasmon polaritons (SPPs) can be excited by light on a metal-dielectric interface. They have shorter wavelength than light at the same given frequency, allowing for control of light propagation at a sub-wavelength length scale.

SPPs in hBN-encapsulated graphene propagate isotropically. When graphene is subject to 2DSL modulation, SPP dispersion relationship depends on the SL gate voltage V_g. At certain values of V_g, only SPP modes along M (or K) direction are allowed.

The fabrication of a graphene 2DSL device for SPP detection is challenging due to the need to make high quality stacks with large area composing of two thin hBN layers and a monolayer graphene sheet. The stacking process can be streamlined by picking up a thick hBN flake with a hole etched away first, and then picking up the thin flakes using the thick hBN flake.

Experimentally we have shown that V_g can indeed change the K vs M direction of SPP propagation in a graphene 2DSL device, acting as a programmable switch. When the shape of the SPP launcher is changed to a needle, we are able to map the SPP band structure equi-V_g contour by taking the Fourier transform of near-field images.

Chapter 6. Future outlook

6.1 Towards a perfectly symmetric 1DSL potential

One of the most exciting promises of a graphene 1DSL system is the ability to induce Landau level degeneracy at CNP thanks to the existence of side Dirac cones at certain strengths of SL modulation u.[1] The most experimentally attainable case is where $u = 6\pi$ and 3 Dirac cones exist at CNP. In a 1DSL modulation potential with both perfect even and odd symmetries, the Landau levels at CNP are expected to be $3 \times 4 = 12$ fold degenerate at $u = 6\pi$, leading to the absence of states with $\sigma_{xy} = \pm 2$. [1] However, as seen in Figure 4.15 the measurement results from my device are inconclusive.

Kang et al [2] has recently suggested creating a 1DSL modulation potential with both perfect even and odd symmetries by using two pairs of periodic local gates with geometries described in Figure 6.1(a). When the gate biases satisfy $V_1 = -V_2$, both even and odd symmetries of 1DSL modulation potential are achieved. According to Kang et al, 1DSL potential induced by patterned dielectric superlattice would always be unable to achieve odd symmetry, therefore gate patterning has to be used.

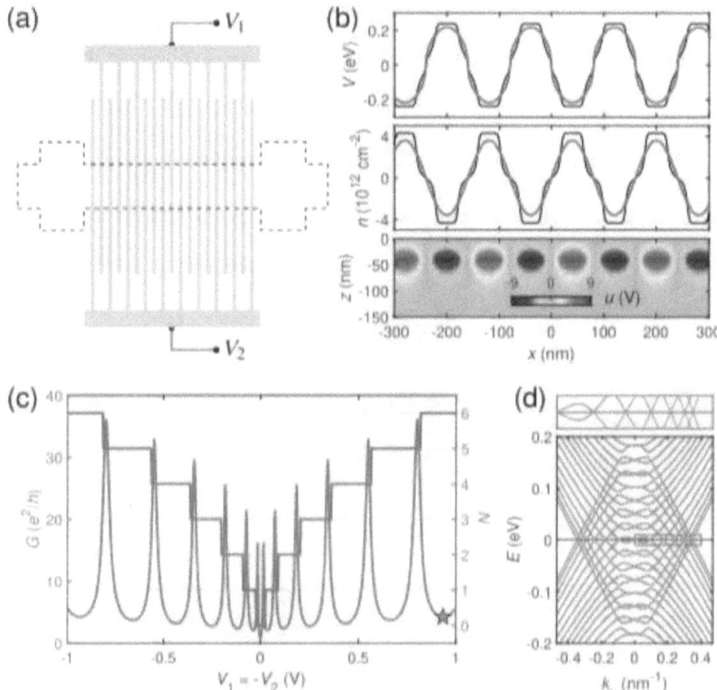

Figure 6.1 Creating a perfectly symmetric 1DSL potential by a pair of local gates

Reprinted from [2]. (a) Schematics of the pair of local gates. Dashed lines indicate the extent

of the Hall bar. (b) Calculated electric potential distribution (bottom), carrier density profiles

(middle), and on-site energy profile (top) for a 1DSL potential with perfect even and odd

symmetries (c) Calculated two-terminal conductance (red) and number of side DP pairs at

CNP (blue) as a function of $V_1 = -V_2$, assuming a 5nm hBN thickness between graphene

and periodic back gate. (d) Band structure of graphene 1DSL corresponding to the star

marked by (c). Up to 6 pairs of side Dirac points at CNP are seen.

6.2 Corbino measurement of graphene under a concentric ring SL potential

Bulk conductance data measured from a graphene device with Corbino geometry has been shown to outperform Hall bar measurements, with improved resolution observed for both the integer and fractional quantum Hall states (Figure 6.2b,c). [3] Therefore it is natural to make and measure a graphene Corbino device with concentric circles etched on SiO_2 substrate as the SL modulation. The lack of any edge transport in Corbino devices could suppress off-angle scattering seen in Hall bar devices and allow ballistic transport at high temperature scales.

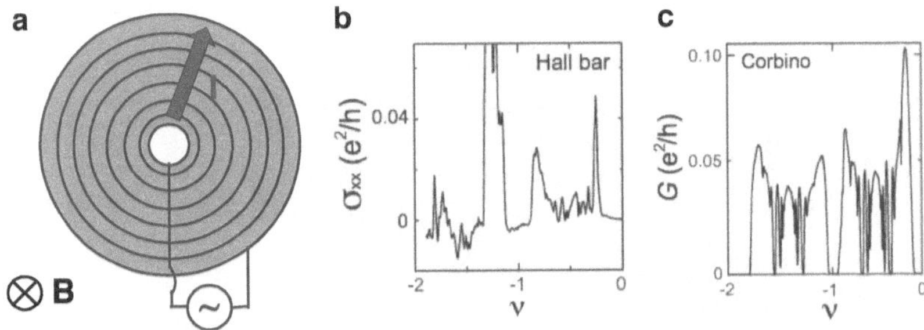

Figure 6.2 Corbino devices (a) Schematics of a graphene Corbino device with concentric rings centered at the center electrode etched on the SiO_2 dielectric substrate. (b) Longitudinal conductivity of a graphene Hall bar device as a function of filling fraction ν measured at B=15T and T=300mK. (c) Bulk conductance of a graphene Corbino device as a function of filling fraction ν measured at B=15T and T=300mK. (b,c) reprinted from [3].

6.3 Combination of patterned superlattice and moiré SL

In magic angle twisted bilayer graphene, flat bands result in many-body effects such as Mott-like insulators[4] and superconductivity[5]. Unlike the flattened bands in graphene 1DSL, flat bands in magic angle tBLG are separated from neighboring bands in energy. It is natural to ask what would happen when we combine these two types of band flattening in one system.

One possibility is that the flat bands in tBLG may fold at the edges of the superlattice Brillouin zone, creating replicated mini-flat bands.

6.4 Graphene subject to a quasicrystal superlattice potential

A quasicrystal is a structure that is ordered but not periodic. Quasicrystal structure occur in tBLG with twist angle 30° where multiple Dirac cones replicated with a 12-fold rotational symmetry exist.[6] However, these cones exist at energies about 1.7eV above CNP, which is too high to be reached by field-effect gating. Figure 6.3 shows the "band structure" of monolayer graphene under an 8-fold quasicrystal potential at two different strengths of SL modulation. The band structure features related to the quasicrystals happen at ~0.1meV above CNP, making it realistic to probe these features by transport measurements.

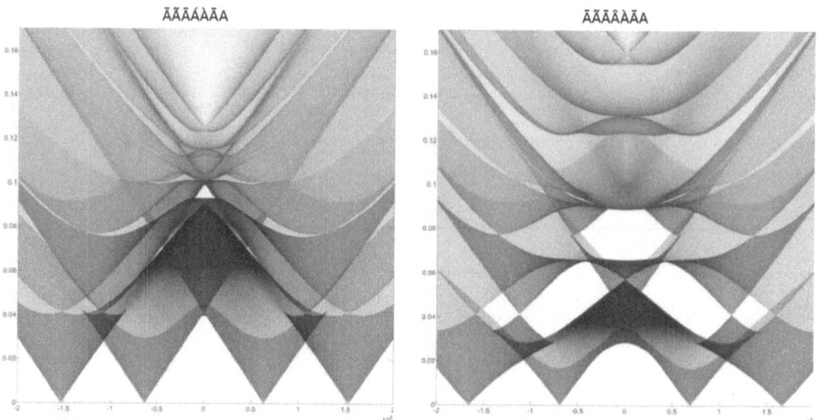

Figure 6.3 "Band" structure of monolayer graphene under and 8-fold quasicrystal potential. Lattice constant=50nm. Energy unit on the vertical axis is meV.

www.ingramcontent.com/pod-product-compliance
Lightning Source LLC
LaVergne TN
LVHW041733190726
843493LV00008B/2332